AF748759

# Reimagining Resources and Community Development

This book analyzes the experiences of communities facing major challenges relating to resource dependency and community sustainability, drawing on specific examples from the Canadian province of Newfoundland and Labrador. It offers a methodology of self-analysis for communities facing similar challenges, inspired by the ups and downs, local strategies for self-analysis, and collaborative work toward new futures in this Canadian province.

Life in hundreds of small coastal settlements revolved around the cod fishery, until the fishery was no more viable. Communities have had to rethink their strengths, reconsider their assets, and imagine potential futures in the wake of events such as colonization and the collapse of the fishing industry. Their experiences are relevant for other parts of the world where formerly central resources are depleted or lose their value, and communities face the need for transition. The capacity to imagine different futures is rooted in the ability to critically consider strengths and weaknesses alike. The authors skillfully dissect and illuminate the conditions that can enable the reconsideration of local assets and narratives, toward a more sustainable future. The variety of these conditions, ranging from social memory to public debate, policy tools and institutional capacity, decision arenas, paths for participation, and distributed strategic leadership, are laid out clearly and illustrated vividly through vignettes written by individuals who participated in the events described. This book culminates in a flexible yet clearly structured method of self-analysis, useful for communities interested in rethinking their strengths and working toward new futures.

This book will appeal to students, scholars, and professionals interested in community development and redevelopment and offers a new understanding of the mechanics of local and regional resilience.

**Monica Gruezmacher** is an Adjunct Professor in the Environmental Policy Institute at Grenfell Campus, Memorial University and in the Department of Earth and Atmospheric Science at the University of Alberta, Canada.

**Kelly Vodden** is a Research Professor in the Environmental Policy Institute at Grenfell Campus, Memorial University, Canada.

**Brennan Lowery** is an Assistant Professor in the Social Entrepreneurship and Non-Profit Management Program at St. Thomas University, Canada.

**Amy Hudson** is the Owner/Director of Husky Consulting Inc., Happy Valley-Goose Bay, Canada and an Adjunct Professor in the Environmental Policy Institute at Memorial University, Canada.

**Kristof Van Assche** is a Professor at the University of Alberta, Canada and a Visiting Professor at the University of the Free State, South Africa.

# Earthscan Studies in Natural Resource Management

**Governing Natural Resources for Sustainable Peace in Africa**
Environmental Justice and Conflict Resolution
*Edited by Obasesam Okoi and Victoria R Nalule*

**Capacity-Building and the Water-Energy-Food Nexus**
Rethinking Integration in the Asia-Pacific
*Maureen Papas*

**Governing the Palm Oil Industry**
Perspectives from Southeast Asia and Latin America
*Edited by Patrick O'Reilly and Helena Varkkey*

**Researching Institutions in Natural Resource Governance**
Methods and Frameworks
*Edited by Fiona Nunan*

**Sustainable Management of Cordyceps**
Supply Chains and Resource Management Policies
*Edited by Jiping Sheng and Ksenia Gerasimova*

**Reimagining Resources and Community Development**
Lessons from Newfoundland and Labrador
*Monica Gruezmacher, Kelly Vodden, Brennan Lowery, Amy Hudson and Kristof Van Assche*

For more information about this series, please visit: https://www.routledge.com/Earthscan-Studies-in-Natural-Resource-Management/book-series/ECNRM

# Reimagining Resources and Community Development

## Lessons from Newfoundland and Labrador

**Monica Gruezmacher, Kelly Vodden, Brennan Lowery, Amy Hudson and Kristof Van Assche**

Routledge
Taylor & Francis Group
LONDON AND NEW YORK

earthscan
from Routledge

First published 2026
by Routledge
4 Park Square, Milton Park, Abingdon, Oxon OX14 4RN

and by Routledge
605 Third Avenue, New York, NY 10158

*Routledge is an imprint of the Taylor & Francis Group, an informa business*

© 2026 Monica Gruezmacher, Kelly Vodden, Brennan Lowery, Amy Hudson and Kristof Van Assche

The right of Monica Gruezmacher, Kelly Vodden, Brennan Lowery, Amy Hudson and Kristof Van Assche to be identified as authors of this work has been asserted in accordance with sections 77 and 78 of the Copyright, Designs and Patents Act 1988.

The Open Access version of this book, available at www.taylorfrancis.com, has been made available under a Creative Commons Attribution-Non Commercial-No Derivatives (CC-BY-NC-ND) 4.0 license.

Any third party material in this book is not included in the OA Creative Commons license, unless indicated otherwise in a credit line to the material. Please direct any permissions enquiries to the original rightsholder.

*Trademark notice*: Product or corporate names may be trademarks or registered trademarks, and are used only for identification and explanation without intent to infringe.

*British Library Cataloguing-in-Publication Data*
A catalogue record for this book is available from the British Library

ISBN: 978-1-041-02632-7 (hbk)
ISBN: 978-1-041-02635-8 (pbk)
ISBN: 978-1-003-62029-7 (ebk)

DOI: 10.4324/9781003620297

Typeset in Times New Roman
by codeMantra

# Contents

# Vignette Authors

**Jasen Benwah** is the Chief of the Benoit First Nation, Cape St George, Newfoundland.

**Karin Buhmann** is a Professor in the Business School, Department of Management, Society and Communication at Copenhagen Business School, Copenhagen, Denmark.

**Michael Burzynski** is a former Parks Canada scientist. Rocky Harbor, Newfoundland.

**Fred Campbell** is a former extension program developer, journalism/communication technology practitioner/instructor, mixed media producer, and PhD candidate. Hantsport, Nova Scotia.

**Ivan Emke** is a facilitator at Kitchen Table Communications, creator and host of the Fit to Eat: The NL Farm and Food Show, and producer at Routes & Branches & Beyond. Kitchener, Ontario.

**Mercy, Goodness, Tabitha, and Ebenezer Fabulous** belong to a Nigerian Family. Goose Bay, Labrador.

**John Fisher** is an entrepreneur and the Owner of the Fishers Loft, Port Rexton, Newfoundland.

**Robert Greenwood** is Deputy Minister of Rural and Regional Development and Engagement and Chief Economic Development Officer for the Province of Newfoundland and Labrador, St. Johns, Newfoundland.

**Amy Hudson** is an Indigenous scholar and adjunct faculty member in the School of Science and the Environment, Grenfell Campus, Memorial University. She is also the Owner and Director of Husky Consulting, Inc., Happy Valley-Goose Bay, Canada.

**Umme Kulsum** is a post-doctoral fellow at Grenfell Campus, Memorial University, Corner Brook, Newfoundland.

**Anne Marceau** is a former Parks Canada scientist. Rocky Harbor, Newfoundland.

**Shirley Montague** is the founder of Trails Tales Tunes Festival, Norris Point, Newfoundland.

**Sue Newhook** is a former CBC Journalist and Journalism Professor. Halifax, Nova Scotia.

**John Norman** is the Mayor of the town of Bonavista and entrepreneur. Bonavista, Newfoundland.

**Roger Pickavance** is a food writer and retired Professor at Memorial University, St. Johns, Newfoundland.

**Edith Samson** is the Director of the Coaker Heritage Foundation, Port Union, Newfoundland.

**Mark C.J. Stoddart** is Professor of Sociology at Memorial University, St. Johns, Newfoundland.

**Nina Tourett-Retieffe** is a member of the Benoit First Nation and Indigenous Community Liaison, Grenfell Campus, Memorial University, Mainland, Newfoundland.

**Avery Velez** is a transdisciplinary Mi'kmaw scholar at Grenfell Campus-Memorial University, Corner Brook, Newfoundland.

**Paul Wylezol** is the Chairperson at International Appalachian Trail and Cabox Aspiring Geopark, Corner Brook, Newfoundland.

# Acknowledgments

We would like to thank all the vignette authors and artists for sharing their thoughts, experiences, and the many different colors of Newfoundland and Labrador.

This book draws on research supported by the Social Sciences and Humanities Research Council of Canada (No. 611-2019-0275 and NFRFT-2020-00452).

# 1 Introduction

## Engaging creatively with past and present assets to forge new futures

**Abstract**

*What is important in life, and what is important to live? Questions like these are implied in many decisions individuals make daily, and they may surface when communities must think about, and (re)make their futures. This book is about communities that have been repeatedly confronted with these questions, and it is about hundreds of small communities dotting a rugged coastline with a history of colonialism and marginalization, and yet also of resilience and self-reliance despite enduring dependence on centers of political and economic power. It is about how communities have responded when key resources become scarce or inaccessible and how livelihoods and cultures persist after being deeply threatened. In such cases, a process of collective rethinking of economy, governance, and identity is needed, a process that has occurred intensely on and off in the Canadian province of Newfoundland and Labrador.*

Community for us is self-defined, a unit sharing an identity and means of collective coordination. In this book, we will be discussing the importance of re-examining community assets and strengths through a process of self-analysis drawing lessons and inspiration from Newfoundland and Labrador (NL). Ultimately, we present an approach that may be applicable in other regions of the world facing similar circumstances.

The process of rethinking *which* assets and strengths are or could be of value for future development and community well-being is always challenging. Livelihoods must be reconsidered, landscapes perhaps used in different ways, and stories about community and environment critically re-evaluated, particularly as the world changes at an increasingly rapid speed. Such an endeavor tends to be difficult for several reasons. One difficulty highlighted throughout this book is that institutions, meaning the rules of coordination

DOI: 10.4324/9781003620297-1
This chapter has been made available under a CC-BY-NC-ND 4.0 license.

within a community and between communities and their environments, that are meant to serve a stabilizing role may cease to do so in places going through dramatic changes (Van Assche et al., 2021). This can make possible paths of change hard to discern and the process of adaptation hard to implement.

A second difficulty relates to the first: community futures often go unquestioned. Local infrastructures to support and encourage critical reflection may be missing, and possible futures often do not easily reveal themselves, especially after a shock. Difficulties can trigger self-doubt, but they can also bring about the reassertion of old values, stories, and identities. There are strengths and limitations to any story, as well as different ways of telling the story. This includes stories of identity, stories that explain the world, stories that link community and environment, and stories that connect past, present, and future. Many stories are tied to cultural self-definition or shared values. Rethinking or questioning such values can feel like heresy or be perceived as oppressive when the rethinking is prompted externally. Rediscovering hidden assets and strengths, or discovering new ones, can be difficult due to entrenched understandings and cherished narratives that create a lens through which a community understands itself and the world around it. Yet as the world changes around us, stories change as well, helping in turn to re-shape community futures.

We offer here an approach for self-analysis that can incorporate a suite of methods and tools to help communities define and organize assets and strengths and thus enable collective strategy where it may have seemed difficult before. The core of this self-analysis is collective reflection, likely with help from both insiders and outsiders, a process of self-assessment for re-evaluating and (re)discovering assets and opportunities. This includes an examination of knots, rigidities, blind spots, and self-limiting aspects for thinking and organizing, but also ways to overcome these obstacles.

This book introduces four infrastructures for community self-reflection. By infrastructures here we mean a collection of different elements that enable the revisiting of assets, strengths, and futures, to which we devote four chapters in this book:

1. *Narratives*, stories, and identity, with a special role for identity narratives that frame assets.
2. *Public discourse and learning*, with a special interest in dialectical learning and diversity of perspectives.
3. *Governance*, with particular attention to governance reforms and experiments for building institutional capacity.
4. *Leadership and participation,* with a specially articulated set of leadership roles where sustainability threats occur.

These distinct but interconnected infrastructures each rely on one another to function. Even when governance capacity exists, and governance tools and goals are considered legitimate in the community, leadership is required that goes beyond managing or stabilizing routines. Even when public discourse offers space for diverse and critical voices, this will have a minimal effect if there is no functional arena for collective decision-making or no tools to translate these decisions into institutions and organizations that are recognized elsewhere and function locally.

While not identified as an infrastructure on its own, we argue that creativity is also a requirement. This assertion aligns with the existing literature and ideas related to innovation and adaptation, in which creative methods and ways of viewing community can be crucial for questioning seemingly hard truths and finding alternative narratives for what may have seemed obvious in the past. Creativity is also needed for strategizing toward new forms of organizing and using space, relating to resources, finding new ways to connect, and representing the community both to itself and to the outside world. We offer these four infrastructures as focal points for rethinking assets and futures, while acknowledging that this is but one of many possible approaches. Just as there are no one-size-fits-all recipes for creativity, nor for rethinking assets and strengths, or for the construction of community futures.

## A brief history of Newfoundland and Labrador community development

The lands and waters now known as the Canadian province of NL, lying in and alongside the North Atlantic, are rich in natural resources, culture, and often astonishing beauty. Yet the communities and regions of this part of the world have faced difficult questions and changes many times over their history. Many of these changes have been related to natural resources and local ecosystems. Diverse peoples and cultures have relied on these resources, which have evolved over time, along with the human uses and relationships with these environments. NL communities have also adapted over time to a continually shifting geopolitical landscape, processes that will be elaborated on further throughout this volume.

The original peoples of NL include the Inuit and Innu of Labrador and the Mi'kmaq and Beothuk of the island of Newfoundland (*Ktaqmkuk*).[1] The Beothuk relied on the formerly large herds of caribou and coastal resources like seal, capelin, smelt, and sea birds (Tuck, 2022). Mi'kmaw peoples were and continue to be supported by and live in relationship with the island's natural resources, including caribou, eels, and salmon, based on the concept of *Netukulimk*, making livelihoods from the land based on relationships of respect, reciprocity, and protection (Barsh, 2002; Prosper et al., 2011). Innu and Inuit peoples also occupied and relied on the resources provided by the lands, waters, and ice of Labrador – and continue to do so (see the excerpt by Hudson below).

Archaeologists have revealed that the Norse (or Vikings) established at least one settlement on the island of Newfoundland in the 11th century (Ingstad & Ingstad, 2000). They were followed by English-Italian John Cabot, who declared 1497 a moment of "discovery," and by European fishers who first visited and later lived along the coasts of NL as the process of European colonization began to unfold (Cadigan, 2009).

From 1713 until the early 20th century, the English fought against the French for the bounty of these lands and waters and to gain access to the territory. Cultural remnants remain of a shifting "French shore" that allocated fishing rights to the French along the Northeast Coast and Northern Peninsula of Newfoundland. An emerging English-Scottish elite concentrated in the town of St. John's, while a growing number of additional settlements appeared along the coastline, including predominantly Irish communities. What these settlements had in common was the domination of cod (*Gadus morhua*) in the colonial economy, a fish population that had already started to shrink in the 19th century under the pressures of overfishing (Figure 1.1).

*Figure 1.1* Cover of a report commissioned by the government in 1925 depicting the wealth of resources of the island. Condon, Michael E. (1925) The fisheries and resources of Newfoundland: "The mine of the sea," national, international and co-operative [book] Centre for Newfoundland Studies – Digitized Books (SH 225 C6).

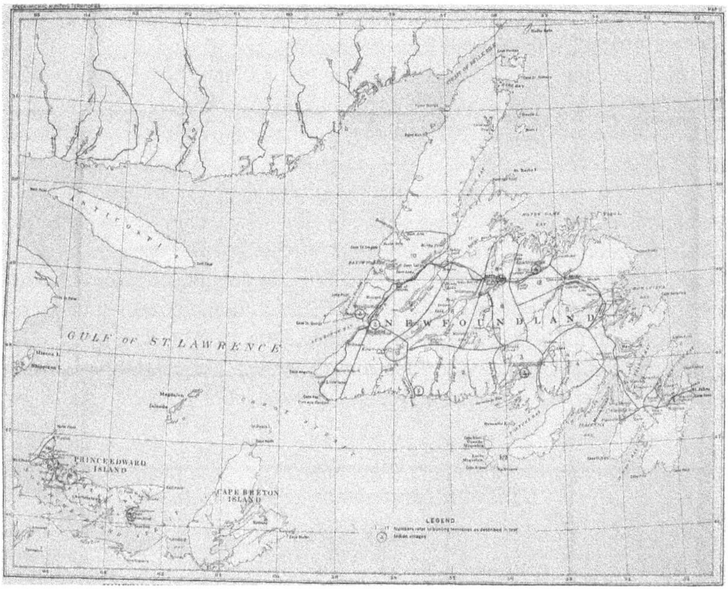


*Figure 1.2* Speck, Frank G., 1922. *Hunting Territories of Micmac Indians in Prince Edward Island and Newfoundland.* [image]. California Digital Libraries, American Digital Libraries (SRLF UCSD:LAGE-139912).

As the cod fishery (as well as salmon and other fisheries) expanded and previously migratory European newcomers began to settle permanently, the powerful merchant houses of Western England began to see this influx of settlers as potential competition (Gillmor, 1980; Mannion, 2018). The colony was not supposed to develop into an entity with its own interests and ideas about fish and its profits; however, informal settlement could not be stopped. St. John's, the capital, only became a municipality in the 19th century but had already started to behave as a colonial metropole, exerting authority over hundreds of smaller settlements where financial capital and external networks, including market access, were scarce.

Within these outport settlements, power was often concentrated in a few hands, and usually not those of local elected officials, as municipalities were rare prior to Confederation. There was a minister or priest, and often a merchant family that owned the main store and had connections to St. John's, or even straight to England; in many cases, merchants were the only ones with enough capital to finance local fishing operations (Cadigan, 2009; Smith, 1968). Fishers borrowed equipment and provisions and had to pay them back at the end of the fishing season with their catch, making them eternal borrowers in a quasi-feudal system in which fishers saw minimal if any profits minimal, if any, profits from their catches.

The Indigenous peoples of what is now the province of NL responded in multiple ways to these colonial forces, adapting to changes while acting to protect their lands, waters, and resources. Colonial governments were formed that developed policies and legislation to protect their own economic interests and aims of territorial expansion. Relations turned violent in many cases, as armed settlers, for example, blocked the Beothuk from accessing salmon runs, seals, and seabird colonies (Adhikari, 2023; Upton, 2008). Despite promises of "peace and friendship" (via the peace and friendship treaty, see Hudson below), as Lieutenant John Cartwright in 1768 described, "killing prevails amongst our People towards the Native Indians..." (Howley, 1915, p. 45; Upton, 2008, p. 74). Indigenous nations retaliated, and tensions continued to escalate, all while devastating diseases broke out among Indigenous populations (Figure 1.2).

---

### A Labrador Inuit history (Southern Inuit)

Amy Hudson

Labrador has been inhabited for millennia, by several groups of nomadic and semi-nomadic people, organizing their livelihoods around different sets of resources. The influence of early colonists on Inuit society in Labrador is pronounced, with the perils still seen and felt today. Early accounts of European–Inuit interaction along the Labrador coast, dating back to the 15th century, illustrate that Inuit were active agents and persistent in their resistance of European whalers, fishers, and traders, whose intent was to exploit the rich marine resources in the region. Basque whalers were active in (Southern) Labrador from 1540 until the early 17th century, when the cod fishery had become the main form of resource exploitation by non-Indigenous people. Inuit in south Labrador were often met with hostility by European visitors who coveted the region's rich marine resources. Colonial records point to conflict-ridden and often violent relationships between Inuit, French, and British in southern Labrador during the 1600s and 1700s, resulting in deaths and, in some cases, Inuit captured as slaves.

Following the Treaty of Paris in 1763, which transferred colonial administration of Labrador from the French to the British, conflict continued between British subjects and Inuit. At this time, European attempts to control trade relations with Inuit involved both the French and British as they competed for trade while making several attempts to enter treaties with Inuit. British officials perceived the Inuit as allies and trading partners of the French, hoping a treaty relationship with Inuit would advance British profit and interests. In 1765, the British were finally successful in making a peace and friendship treaty with Inuit in

*Figure 1.3* Moravian settlement. Unknown (pre-1919). Okak. Moravian settlement, pre-1919 [still image]. Geography collection of historical photographs of Newfoundland and Labrador (22.03.001: Okak.).

southern Labrador, witnessed by hundreds of Inuit who traveled long distances by sea to attend. The treaty was renewed through exchanges of gifts for several years following, but the treaty didn't end the conflict.

The so-called Moravian missions were among the Inuit encounters with the West that left deep traces in Inuit governance and identity. Moravians are protestant Christians, mostly German-speaking, originally from the current Czech Republic. With German support, the Moravians became zealous missionaries in the early 18th century, settling in Western Greenland in 1733 and northern Labrador by 1771. The Moravians were supported by the British government, who believed them to be a civilizing influence. This included encouraging the Inuit to become sedentary by supporting settlement and bestowing land grants. This policy of isolating Inuit from Europeans and other Inuit groups was intended, at least in part, to reduce the violence (i.e., Inuit resistance) and make for smoother exploitation of resources for the British and the Moravians themselves. The British attempted to keep Inuit north of Chateau Bay and away from French traders in southern Labrador and the Northern Peninsula. Despite efforts toward the containment of Inuit and the pursuit of control and economic gain from a resource-rich land, Inuit would continue to live in the south and along the entire coast of Labrador from the Belle Isle Strait to Hudson's Strait (Edited excerpt from Hudson, 2020) (Figure 1.3).

---

A decline in the productivity of cod, improved canning technologies, and the decline of lobster fisheries elsewhere in North America and Europe led settlers to start fishing lobster, along with salmon and herring (Korneski, 2012). Lobster and sealing saw short-lived booms in the early 20th century but did not replace the all-important cod fishery. Other economic activities like hunting, trapping, and gathering became important to settlers (Hoggart, 1979; Mayda, 2004), and mining and forestry industries began in pre-industrial form before the 1900s.

NL joined Canada in 1949, led by Premier Joey Smallwood. Smallwood's administration sought to reform the fishery and life in small fishing towns, introducing a series of policies and plans to "drag" the province into his vision of modernity (House, 2019; Matthews, 1979). Confederation also continued to disregard the autonomy, rights, and peoplehood of Indigenous groups in the new province, with no mention of Indigenous peoples in the Terms of Union

*Figure 1.4* Joey Smallwood, Newfoundland and Labrador's first Premier with a spade at the opening of a pulp and paper plant. Author Unknown (1971) 00005.04.013–J.R. Smallwood, Economic Development–Melville Pulp and Paper Ltd., Stephenville [still image] Smallwood, Joseph R. – Photographs (00005.04.013).

nor clarification of their status within the new province and country. Instead, Smallwood declared that there were no Indigenous peoples in the province, and it would be years later before they were "recognized" and granted access to services and supports available to other Indigenous peoples in Canada (Hanrahan, 2003). The consequences of this omission are still being felt today (Figure 1.4).

In the decades after joining Canada and the shift to an industrialized frozen cod export fishery, dramatic shifts occurred in outport lifestyles. The term "underdevelopment" was often used to discuss the dire economic conditions of the new province in the mid of Smallwood's "develop or perish" mantra (Côté & Pottie-Sherman, 2020; Crosbie, 1956). Industrial ventures came and went, infrastructure was developed, and several hundred villages, or "outports," were abandoned or moved under government-sponsored resettlement programs (Hoggart, 1979; Withers, 2016). Relocations in northern Labrador also occurred, with dramatic implications for those whose lives and lifestyles were uprooted (Burns, 2006; *Ceremony to Mark Forced Relocation of Inuit Village*, 2012).

Smallwood lost his leadership role in 1972. Subsequent governments softened their industrial policy, and foreign competition in fisheries was diminished by extending the territorial waters of Canada to 200 miles beyond the shoreline, but this did not bring the prosperity that many hoped for when joining Canada. Nor did it solve the ecological problems that were devastating the cod fishery, as foreign overfishing was coupled with Newfoundlanders investing in larger boats, often with provincial and federal government help. These forces ultimately led to a federal fishing ban, a shock from which the province is still recovering.

It is in this diverse and persistent colonial context that communities in NL have been forced to reimagine their futures over time, to adapt and evolve – from early European settlement and intrusion on Indigenous lifeways, to Confederation and the resettlement era, to the groundfish moratoria of the 1990s, and contemporary transitions. If cod was the main asset that drove the development of NL's economy for almost five centuries (at least from a settler and commercial point of view), it had become abundantly clear that relying on the cod fisheries alone could not last.

## Drawing on assets to explore possible futures

Before exploring how to look differently at assets, we must consider what an asset might be. An asset *becomes* an asset for someone or a group of people, in a particular place and time, where something, either tangible or intangible, is considered to be of value to individuals and/or communities with a particular view of the future in mind and within a particular environment (Gruezmacher & Van Assche, 2022). In some cases, the term strength is preferred over asset given the dominant economic valuation associated with the term. In this book, we focus on assets as anything considered of value (e.g., spiritual, cultural, social, ecological, economic, or aesthetic) for a community, based on their valuation of that entity's importance to community members. Assets are therefore contextual, meaning that something can be an asset or a strength in one location or among one group of people at a particular time, but not elsewhere (Van Assche et al., 2024).

We pay special attention to situations where an asset may cease to be an asset, where it cannot function anymore as something the community can use to sustain itself. We recognize that an asset can lose its value in shifting economies and societies or even become dangerous (due to over-reliance on that asset, for example). Moreover, assets look different depending on perspectives on the future; if we are in fear for individual or collective survival, then what was once seen as an asset may look like a threat. If, on the other hand, we believe a boom is coming in the regional economy, a different set of things start to look like assets. Rethinking community futures then most likely will make us look differently at the assets, or strengths, in our communities, allowing us to scrutinize the community's economy, the environment, and forms of organization, and make us reconsider which assets can still be counted on or what other things might suddenly appear as valuable.

Stories are powerful tools for uncovering assets. Stories about place tell us about what people value and the narratives in which they see themselves (Lowery et al., 2020). Even if an international market recognizes something as valuable, this does not mean it is recognized locally as an asset; alternatively, what may be cherished by a community can be entirely overlooked by external markets, government agencies, or the broader societal culture. The stories that define something as an asset can be very different and far removed from the economic sphere. Where a community has existed for a long time, perhaps evolving around a single asset or activity, there's a good chance that what is conceived of as an asset also touches on identity. This is particularly true for Indigenous peoples and communities with close connections to the land and water. Identity is in turn closely connected to emotions that can be evoked like fear, anxiety, and anger, but also happiness and hope.

In places where communities have endured very difficult periods, it can be traumatizing to recall those times or hint at a new period of instability. Such discussions can remind people of the difficulties of rebuilding themselves after a troubled time. New situations can echo old stories, bring back old emotions, or lead to an affective blockage of real discussion of options for the future. Community trauma is not always straightforward to outsiders, and even locals themselves may not always be aware of its implications.

For any community, especially those that have experienced trauma, it is therefore essential to understand that reflection on assets – past, current, and future – is much more than a technical exercise. In many places, layers of impediments may be encountered in reconsidering assets and futures. Hasty decisions like following so-called "best practices" from elsewhere, jumping on a seemingly amazing opportunity, or proposing a clear-cut policy transfer make little sense under such circumstances. Simply bringing people together to discuss and deliberate is likely not enough either, although it is often an important starting point and key component of the reflection and reconsideration process as it unfolds.

## Standard solutions?

Along with others who advocate for place-based approaches to community development, we recognize that standard solutions for community development must be taken with a grain of salt (Ryser et al., 2023; Slawinski et al., 2023). Therefore, we keep a respectful distance from most of the recipes one can find in the literature. Many of these can be valuable, but this entirely depends on the context and the choices made by the community. Participation, local knowledge, and self-organization are all recognized as important elements of planning and acting toward sustainable community futures, particularly within a place-based community development framework (Walzer et al., 2020). Yet, none of these elements can be presented as a silver bullet.

One of the approaches that inspired this book, but should again not be reduced to a recipe, is asset-based community development (ABCD) (Russell, 2022). This approach and associated methods initially appeared in largely underprivileged urban settings in the US, places that were neglected by government and industry alike or feeling the heavy hand of modernist urban renewal. Community organizers, including staff of community development associations and community members themselves, tried to shift the perspective of developers, businesses, city planners, and councilors, revealing the strengths of often poor neighborhoods that may not have been recognized before.

ABCD, as originally developed by Kretzmann and McKnight (1993), was formalized to give communities a process for self-organization and an alternative based in finding resources from within the community rather than looking to investment from elsewhere. It recognized three distinct categories of assets, including individuals (possessing intellectual, practical, and social skills), associations (informal groups, often volunteers), and institutions (understood as more formal structures such as a school, government, or business). ABCD has often been blended with the thinking of community capitals, inspired by American and French sociology, especially Pierre Bourdieu and his notions of social, economic, political, and cultural capitals (Bourdieu, 2011), and the subsequent community capital framework (Emery & Flora, 2006).

While the clear structure of ABCD can be helpful, community development theory and practice show that a productive hybridization can occur between ABCD and other understandings of community development. In theory, the broader community development thinking on participatory development, co-creation, and more recently social innovation and place-based development, meshes with the more flexible understandings of ABCD, emphasizing the value of social networks, local knowledge, self-organization, learning, and the often-hidden diversity of community assets and strengths.

Another approach that both incorporates and serves as an inspiration for ABCD is the livelihoods framework, which centers on thinking about assets in a participatory manner. Livelihoods thinking pays more attention to the ecological environment, viewing livelihoods as ways to use and navigate that environment (Ashley et al., 1999; Scoones, 1998). Depending on the collection of livelihoods that represents local life and culture, even small ecological changes can have large social and economic effects. However, a livelihood is never just a matter of an environment offering an opportunity. What becomes an opportunity does so for a particular community, using certain technologies and human capital, aspiring to certain ideals, sharing certain values, and structured in a particular way.

Therefore, it makes sense to speak of social-ecological systems, where a community and its environment are understood as one, with different interacting subsystems that depend not only on features of the environment but also on the human community(ies) that live within it (Berkes & Folke, 1998; Ostrom, 2009). Food systems, as described by Pickavance below for example, are one such subsystem, illustrating the intimate relationships that exist between people and the environments they are part of. Only when we look at the way the overarching social-ecological system is structured, the way people shape the environment and vice versa, can we recognize why certain livelihoods exist, and why and how something becomes an asset. Not all livelihoods are place-based, and markets are helpful in getting products from far away or selling them to places far away, as well as coordinating dem and and supply. Yet, understanding livelihoods as part of social-ecological systems helps show what becomes a demand, what is valued, and the environments that fulfill this demand, which might not be able to do so forever, particularly without care and responsible, respectful human–environment relationships.

---

**Food in Newfoundland: a personal perspective Community and environment interactions and hidden assets**

Roger Pickavance

I arrived in Newfoundland profoundly ignorant of what food to expect in my new home. But I was enthusiastic about finding out – my university years had developed the latent foodie in me. I traveled around the Island, poking my nose into kitchens and asking questions. People told me about the food they had grown up with, and how they used to catch, gather, or grow what they ate. They described the trials and tribulations of cooking for a family of a dozen or more when water was brought in pails from the well, and you had to cope with the vagaries of a cranky, wood-fired cookstove.

Soon I started writing things down for later translation. One elderly lady told me to bake a cake for "about 10 and a half glasses." Only after she produced a little sandglass was it clear what she meant. But "a handful," "a mugful," "a good bit," or "about half of that old China jug over there" were more problematic. And they were very patient with me when I looked totally confused by "the size of a turr's egg."

I collected hundreds of accounts of traditional ways of life, kitchens, and recipes that spurred me to write my first book, *The Traditional Newfoundland Kitchen*. Then a friend of mine lent me her grandmother's handwritten kitchen notebook. This first-hand look at middle-class St. John's cooking in the early 20th century led to my second book, *Agnes Ayre's Notebook, Recipes from Old St. John's*. Then came *From Rum to Rhubarb* and *From Codfish to Kippers*, which reflects how I was able to hone my love of fish on this Island.

Throughout all the talking, cooking, and research for those books, one repeated theme was that NL has some of the finest ingredients I had ever encountered. To capitalize on them, gastronomy could easily be woven into our existing tourism industry, based on the attractions of the natural world and the intangible appeal of the local people and their way of life. So what should we offer? Above all, we must offer authenticity – difficult to define, but a quality much desired by gastrotourists. What could be more authentic than salt-cod? From the 16th century onward, nothing is more central to the story of Europeans in NL. Salt-cod is usually considered too ordinary, too mundane, to serve to visitors. But we should not be so modest. Just look at the range of salt-cod dishes to be found in France or Portugal.

Our cold-water fish are of superior quality. But for those future gastrotourists, we will have to rise above the choices of deep-fried or pan-fried. Shellfish are highly regarded everywhere, and the quality of shellfish here is second to none. They should be star attractions. Lobsters, for example, are perceived as a luxury food. Our blueberries, partridgeberries (lingonberries), and bakeapples (cloudberries) have helped keep people alive here for centuries. A range of vegetables are traditionally grown, from the cabbage family to leeks and garlic, beets, carrots, and parsnips. All we need are original and interesting recipes, and indeed these are increasingly common, making Newfoundland a foodies' destination.

The kitchens of NL have tremendous potential. Our top-notch ingredients are authentic, tightly linked to landscape, lifestyle, and history. We can reinvent and adapt traditional dishes, combined with less familiar ingredients to foster culinary creativity and appeal, creating dining

*Figure 1.5* Caplin. Credit: Anja Sajovic.

experiences in rural and urban locations that offer an experience of the place, its people, and its food (Figure 1.5).

## ABCD in Newfoundland and Labrador

The 1930s crisis and cooperative movement, followed by Confederation with Canada in 1949, led to a massive asset mapping and visioning effort for the province, accelerated by the expansion of Memorial University College into Memorial University and later its Extension Service, as well as units like the Institute for Social and Economic Research and the Centre for Newfoundland Studies. Throughout the 1950s, specialized government committees inventoried everything that could be considered an asset, contributing to a multitude of plans and visioning exercises in a uniquely intense period of reflection and visioning for the future. But these efforts were often top-down, overlooking local perspectives and visioning power. Related reports featured strengths that were often identified by external experts through central planning efforts. Many pointed to the weakness of local governments as an excuse to keep it that way, keeping local communities dependent on professionals, such as clergy, teachers, merchants, and provincial and federal government officials.

The 1960s and 1970s saw an increase in community-driven development efforts and the growth of local and Indigenous governments and organizations. Rural or Regional Development Associations (RDAs) were grassroots institutions formed for communities to assess their circumstances and pursue development opportunities. Following the recession of the 1980s and then the groundfish moratoria, new governance experiments emerged at local and regional levels, with renewed attention to self-organization, more diverse and locally defined futures, and a broader idea of assets (Hall et al., 2017; Vodden & Hall, 2013).

Provincially, a system of Community Accounts was developed to provide a holistic framework encompassing different aspects of community well-being (Hollett et al., 2008). At the regional level, a sustainability indicators assessment project was undertaken in the Clarenville-Bonavista region (Holisko & Vodden, 2015; Lowery et al., 2019), along with an ambitious Western Newfoundland-Southern Labrador Asset mapping study (Parill et al., 2014). In Gros Morne National Park, a tourism asset inventory and strategic tourism plan was created, building on the Gros Morne Cultural Blueprint in 2011 and the STAR program asset inventory in 2016 (Ginder Consulting & ArtExpert.ca, 2011; Tourism Synergy Ltd. & Broad Reach Strategies, 2016). In the Corner Brook/Gros Morne region, an earlier project had looked for model forest indicators, reflecting on what current and future assets could be, while federal

*Figure 1.6* Cover of booklet created by Black Tickle residents to identify and build on community assets, including photos, quotes, poems, and stories reflecting what contributors love about their community. Credit: Amy Hudson.

requirements led to the creation of Integrated Community Sustainability Plans across the province.

Other efforts at the community (rather than regional) level have included an asset mapping project in Trepassey on the Avalon Peninsula, as well as cultural heritage resources inventories in the communities of Branch and the Fogo Island community of Tilting (St. Croix, 2015). Heritage NL expanded the process of mapping heritage assets in communities through their "People, Places and Culture" workshops (https://heritagenl.ca/programs/ppc/).

In southern Labrador, NunatuKavut Community Council (NCC) supported Southern Inuit in creating a stronger future for their communities through community-driven discussions around sustainability and the Community Governance and Sustainability Initiative. Three pilot communities – Black Tickle, Norman Bay, and St. Lewis – participated in community asset mapping, engagement strategies, visioning exercises, and community initiative planning training, focusing on what is already strong and successful in communities, rather than being defined by problems, and how to build on these strengths (Hudson & Vodden, 2020) (Figure 1.6).

An analysis of these various asset mapping and more general ABCD experiments in the province by Lowery and Vodden (2019) shows the productive hybridization of the ABCD approach with other community development processes. Mapping was sometimes combined with participatory visioning and voluntary regionalization efforts, and was used across the province to mobilize notions of social and community capital. In related work, Lowery et al. (2020) creatively built on the narrative tradition of ABCD in the province and elsewhere. Initial versions of ABCD were open to the use of stories, yet, often mapping was assumed to be a more direct and rationalistic process of naming and listing. While this can still result in a process of discovery as people are connected in new ways, highlighting the importance of stories helps to reveal the complexity of asset construction in a community,

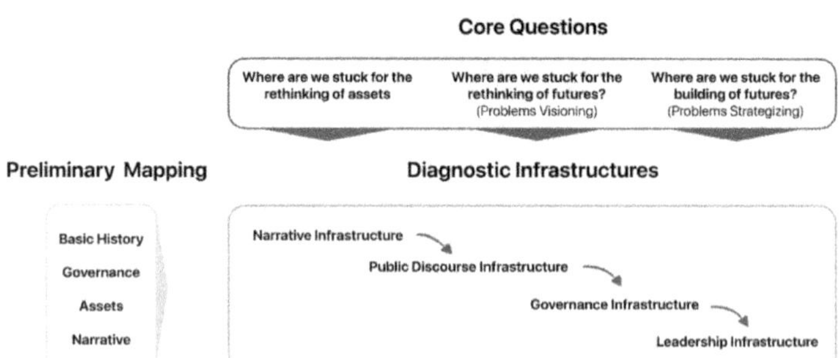


*Figure 1.7* Diagnosing issues in the main infrastructures necessary for rethinking assets and futures. Ideas further developed in Chapter 6.

shared assumptions, insider/outsider dynamics, and cultural continuities and discontinuities.

We believe that the varied stories presented in this volume can be useful for other communities challenged with rethinking their futures, but this cannot lead to universal recipes. Neither ABCD, social-ecological perspectives, nor livelihoods thinking can be used in a formulaic way, but rather as an inspiration. These ideas have helped us reflect on the experiences of communities (and our own experiences) and identify themes and mechanisms, recurring problems, and opportunities, that can help communities in NL and elsewhere in navigating the formidable task of rethinking what might be of value for the future and in what ways.

## Methods and tools

The key approach this book offers is based in mapping, understanding, and improving the infrastructures for rethinking assets, with the aim of providing new flexibility in thinking and acting as a community. In the process, obstacles and enablers can be found in these different infrastructures. We suggest different possible methods (a selection found in the annex) that can be employed in an ideally collective reflection on these infrastructures. The umbrella approach offered here must be consciously specified and adapted by communities themselves, yielding outcomes that are unique to each community (see Figure 1.7).

We believe that for understanding all four infrastructures, it is first useful to understand the history of the community and its governance system, even where governance may be largely informal. Furthermore, mapping of assets or strengths and narratives ought to be linked from the beginning (in preliminary mapping). Mapping stories is limited by practicalities, but it is possible to recognize in conversation the key identity narratives, narratives about past and future (key periods, moments, shifts, hopes), stories about problems, the environment, and community values and governance (Brenneis, 1988; Sools et al., 2015). For the (preliminary) mapping of assets, one can identify assets of the past, present, and (anticipated) future, assets that are currently recognized, not recognized enough, or perhaps over- or underused. Finally, one can identify features that might not be assets at all, but in contrast have been creating problems. Linking again to governance, considering what is an asset or strength must also come in concert with discussions on what a desirable future might look like.

---

### Assessing community well-being according to *Kloqowej* and the four directions

Avery Velez

Indicators used to assess community well-being and sustainability are often based on breaking well-being into distinct measures or categories of whether a community is doing "well" according to pre-determined notions. Yet, well-being and sustainability are not one-size-fits-all concepts, and perceptions of well-being and sustainability depend heavily on the worldviews of those defining them. For example, Indigenous Services Canada's Community Well-Being Index relies on statistics surrounding education, labor force activity, income, and housing to evaluate the well-being of Indigenous communities. These indicators reveal the worldview and biases behind their development and are based on a definition of well-being and sustainability that is incompatible with many Indigenous nations' priorities. Specifically, many of these indicators rely upon engagement with the colonial state and focus on western education, engagement in the colonial capitalist labor market, value based on wealth, and western measures of social organization via sedentary settlements, none of which necessarily support Indigenous understandings of our own communities' and nations' well-being and sustainability.

In comparison, some more holistic Indigenous interpretations of well-being and sustainability include diverse indicators that cover a broader range of experiences that either directly challenge or live beyond colonial constructs. I propose consideration of an Indigenous framework for well-being based on the *kloqowej*, the Mi'kmaw eight-pointed star.

The *kloqowej* is a symbol for the Wapna'ki people, the Indigenous nations, including Mi'kmaq, in what is now known as Atlantic Canada, northern Maine, and the Gaspé peninsula. It represents the connections of everything and everyone in the land. Like the pan-Indigenous medicine wheel, it points in four directions, each of which have their own interrelated meanings. Mi'kmaw legal scholar and Elder Tuma Young writes about the *kloqowej* as a model for Mi'kmaw law and justice, but it also provides a useful framework for community well-being and sustainability. The directions represent the following:

1 East: understood as the source of peace. In the L'nu [Mi'kmaw] sense of the word, it is the fundamental principle regulating the establishment, development, and maintenance of respectful, supportive relationships between individuals, families, communities (human and nonhuman), and nations.
2 South: where kindness resides, providing a "guide to right behavior" in Mi'kmaw people's interpersonal interactions.
3 West: holds the principle of sharing and an understanding of property rights not in the sense of exclusive "private ownership of land" but emphasizing collective rights and responsibilities to the land carried out for the best interests of future generations.
4 North: the direction where trust is found, the foundation for maintaining relationships in a good way, including relationships based on Treaties.

These four directional principles reveal a conceptualization of community well-being and sustainability based in relationships between Mi'kmaq and *nmitkinen aqq msit ko'kmanaq* (our territory and all our relations) that are peaceful, harmonious, kind, and based on sharing,

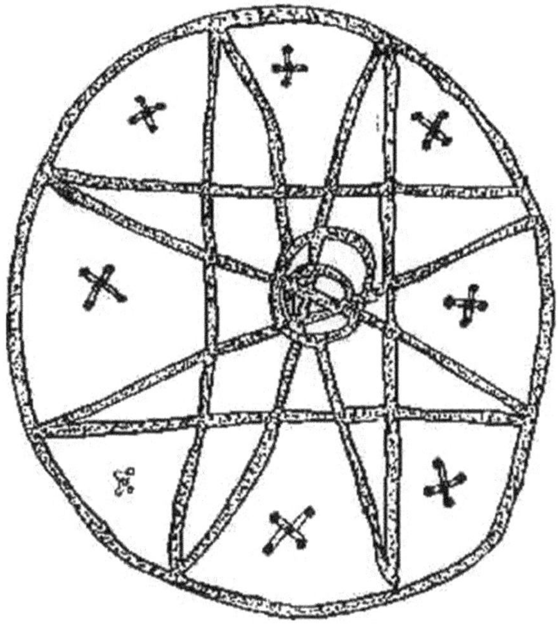

*Figure 1.8* Illustration of Mi'kmaw eight-pointed star n.d., pre-1500 A.D. [image petroglyph] Mylar tracing: R.H. Whitehead, 1983. Nova Scotia Museum Halifax (P179/N-17,124).

respect, reciprocity, and responsibility for the future. Although the development of a well-being framework based on the *kloqowej* model requires further thought, it presents a potential innovative approach to evaluating well-being and sustainability based on the core Mi'kmaw values and principles rather than on often dominant colonial capitalist principles of wage-labor, western education, and profit margins (Figure 1.8).

---

This brief quote from a recent newcomer to Labrador illustrates what an outsider has learned to appreciate in the community.

> As we do know that Canada is one of the destinations for the immigrants, with people from all walks of life. My family and I loved to live in Canada right from Nigeria before we moved to the United Kingdom. We love to explore new places and then relocated to Happy Valley Goose Bay a few weeks ago. The town of Goose Bay is full of tradition and culture. It is an example of a classic town with polite and "happy to help" people.
>
> Furthermore, there are different kinds of organizations that provide countless services to newcomers like Association for New Canadians

> (ANC); which is incredibly helpful and supportive to everyone especially to the newcomers. They don't only provide the needed information but ensure it is achievable. ANC organizes events that makes one feel at home, meet new people and stay connected.
>
> Mercy, Goodness, Tabitha, and Ebenezer Fabulous
> (Nigerian family in Goose Bay, Labrador)

Throughout the book, we encourage consideration of *different perspectives*, from within and outside the community, and active efforts to learn from differences. Different perspectives can come from perhaps unexpected insiders, such as children, elderly people, or newcomers. They can appear in discussions, but also in artistic expressions (Gérin & McLean, 2009; Kapitan et al., 2011). Community art can be helpful as a focus of public discussion, even if some controversy is involved (Van Der Vaart et al., 2018).

---

**Understanding community strengths through children's eyes: How unexpected insiders can bring to light a different perspective on what is of value in the community**

Nina Tourett-Retieffe and Chief Jasen Benwah, Benoit First Nation

*On May 24, 2023, grades 4, 5, and 6 children of Our Lady of the Cape School in Cape St. George, NL were asked what is it that they love about their community and to express their thoughts through watercolors and short written descriptions.* The activity was meant to transcend the classroom and spark conversation among other community members on what they appreciate about the place they live in. The artwork revealed histories and stories about the past and present of the community, unique connections that hopefully inspire an ongoing reflection.*

*Figure 1.9* Illustration of children's artworks depicting some of the things they love about their community. Artwork by George Felix age 10 and Charlie age 10.

In the depths of their innocence, children can teach us intelligence. Their unconscious knowledge is invigorating. As they express themselves with eagerness, their perceptions are somewhat magical. Can they teach us to stop and see what strengths we truly have?

They see what we refuse to see. The colors and tranquility in their work are vibrant. Their youthful minds are sponges that see the beauty within our community. With the natural paint pigments, and the stroke of their brushes, their community comes alive on the sheets of plain white paper as they tell us what we think they don't see.

From our beautiful ocean, forest, animals, and vegetation, the children capture what they love about their community. The color changes from night to day, the waves of our magnificent ocean, and the details of the beaches. This art-based project helps us understand community strengths through children's eyes (Figure 1.9).

**The activity received the support of researchers from Memorial University, University of Alberta, and University of Waterloo through a grant from the Social Sciences and Humanities Research Council of Canada.*

This book intends to show how the obstacles for informed reconsideration of assets are indeed layered and entangled. Many things are needed to provide communities with the freedom to look anew at their usual practices and stories and rethink assets. All four infrastructures might be at play. Analyzing their entanglement can be helpful, not just in diagnosing rigidities in the rethinking of assets but also in pointing toward solutions. Those solutions, we argue, often ask for a community strategy, while such strategy can take many shapes.

Throughout the book, we emphasize that standard solutions and best practices ought to be approached with a degree of skepticism and that solutions are most likely to emerge locally. Assets are contextual. They emerge from a perspective on the future that must be recognized in and by the community. Thus, the procedures to envision and rethink assets must also be contextual. Not all communities are ready for the same questions at the same time or for the same modes of self-analysis. For this reason, thinking of assets should be an ongoing process, reflecting changes that take place in the community over time and drawing from approaches to assessing community strengths and well-being that work best for each individual community or regional context at a particular point in time.

Telling the story of NL communities in its entirety is not possible in the confines of one book, or even necessary for our purposes. What is more interesting for us, and we hope for audiences in many places struggling to redefine themselves, is to share some of the problems and solutions that have emerged from this place and its people. The main narrative and the local voices and experiences present in the vignettes regularly interspersing it, share lessons about what is important for a community or region when confronted with the need to do or become something new.

## Note

1 While the Beothuk are no longer considered to exist as a culturally and politically distinct people increasingly recognized, Mi'kmaw oral histories suggest the Beothuk intermarried with other Indigenous nations rather than having gone "extinct" (Tuck, 2022).

## References

20th-Century Fisheries and Settlement Patterns. (n.d.). Heritage Newfoundland and Labrador. Retrieved April 10, 2024, from https://www.heritage.nf.ca/articles/economy/20th-century-fishery.php

Adhikari, M. (2023). 6 "Now We Are Natives": The genocide of the Beothuk people and the politics of "Extinction" in Newfoundland. In F. Jacob & K. S. Todzi (Eds.), *Genocidal violence: Concepts, forms, impact* (pp. 115–136). De Gruyter Oldenbourg.

Ashley, C., Carney, D., & Britain, G. (1999). *Sustainable livelihoods: Lessons from early experience* (Vol. 7, Issue 1). Department for International Development.

Barsh, R. L. (2002). Netukulimk past and present: Míkmaw ethics and the Atlantic fishery. *Journal of Canadian Studies*, *37*(1), 15–42.

Berkes, F., & Folke, C. (1998). Linking social and ecological systems for resilience and sustainability. In F. Berkes & C. Folke (Eds.), *Linking social and ecological systems: Management practices and social mechanisms for building resilience* (pp. 1–25). Cambridge University Press.

Bourdieu, P. (2011). The forms of capital. In Imre Szeman and Timothy Kaposy (Eds.), *The sociology of economic life* (3rd ed.) (pp. 81–93). Routledge.

Brenneis, D. (1988). Telling troubles: Narrative, conflict and experience. *Anthropological Linguistics*, *30*, 279–291.

Burns, A. (2006). Moving and moving forward: Mushuau Innu relocation from Davis Inlet to Natuashish. *Acadiensis, 35*, 64–64.

Cadigan, S. T. (2009). *Newfoundland and Labrador: A history.* University of Toronto Press.

Ceremony to Mark Forced Relocation of Inuit Village. (2012, August 15). *CBC News.* https://www.cbc.ca/news/canada/newfoundland-labrador/ceremony-to-mark-forced-relocation-of-inuit-village-1.1163442

Côté, I., & Pottie-Sherman, Y. (2020). *Resettlement: Uprooting and rebuilding communities in Newfoundland and Labrador and beyond.* Memorial University Press.

Crosbie, J. C. (1956). Local government in Newfoundland. *Canadian Journal of Economics and Political Science/Revue Canadienne de Economiques et Science Politique*, *22*(3), 332–346.

Emery, M., & Flora, C. (2006). Spiraling-Up: Mapping community transformation with community capitals framework. *Community Development*, *37*(1), 19–35.

Gérin, A., & McLean, J. S. (2009). *Public art in Canada: Critical perspectives.* University of Toronto Press.

Gillmor, D. A. (1980). Newfoundland: Resources and development. *Studies: An Irish Quarterly Review*, *69*(275/276), 253–259.

Ginder Consulting, & ArtExpert.ca. (2011). Gros Morne cultural blueprint. https://artexpert.ca/wp-content/uploads/2020/08/ArtExpert_Gros-Morne-Cultural-Blueprint.pdf

Gruezmacher, M., & Van Assche, K. (2022). *Crafting strategies for sustainable local development. A Community Guide*. In Planning, Groningen.

Hall, H., Vodden, K., & Greenwood, R. (2017). From dysfunctional to destitute: The governance of regional development in Newfoundland and Labrador. *International Planning Studies*, *22*(2), 49–67.

Hanrahan, M. (2003). The lasting breach: The omission of aboriginal people from the terms of union between Newfoundland and Canada and its ongoing impacts. Royal Commission on Renewing and Strengthening Our Place in Canada. http://www.exec.gov.nl.ca/royalcomm/research/pdf/Hanrahan.pdf

Hoggart, K. (1979). Resettlement in Newfoundland. *Geography*, *64*(3), 215–218.

Holisko, S., & Vodden, K. (2015). *Assessing the factors impacting the sustainability of the Clarenville-Bonavista Rural Secretariat Region: Phase two final report*. Grenfell Campus, Memorial University of Newfoundand.

Hollett, A., May, D., & Giles, C. (2008). The Newfoundland and Labrador system of community accounts. In *Statistics, knowledge and policy 2007: Measuring and fostering the progress of societies* (pp. 481–492). OECD, SPO, Turkstat.

House, D. (2019). Did smallwood neglect the fisheries? *Newfoundland and Labrador Studies*, *33*(2), 436–458.

Howley, J. P. (1915). *The Beothucks or Red Indians: The aboriginal inhabitants of Newfoundland*. Cambridge University Press; Memorial University of Newfoundland. Libraries. Centre for Newfoundland Studies.

Hudson, A. (2020). *Reclaiming Inuit governance: Planning for a sustainable future in Nunatakavut* [Doctoral, Memorial University of Newfoundland]. https://research.library.mun.ca/14950/

Hudson, A., & Vodden, K. (2020). Decolonizing pathways to sustainability: Lessons learned from three Inuit communities in NunatuKavut, Canada. *Sustainability*, *12*(11), 4419.

Ingstad, H., & Ingstad, A. S. (2000). *The viking Discovery of America: The excavation of a norse settlement in L'Anse Aux Meadows, Newfoundland*. Breakwater Books.

Kapitan, L., Litell, M., & Torres, A. (2011). Creative art therapy in a community's participatory research and social transformation. *Art Therapy*, *28*(2), 64–73.

Korneski, K. (2012). Development and degradation: The emergence and collapse of the Lobster Fishery on Newfoundland's West Coast, 1856–1924. *Acadiensis*, *41*, 21–48.

Kretzmann, J. P., & McKnight, J. (1993). *Building communities from the inside out: A path toward finding and mobilizing a community's assets* (p. 376). Asset-Based Community Development Institute, Institute for Policy Research, Northwestern University.

Lowery, B., Dagevos, J., Chuenpagdee, R., & Vodden, K. (2020). Storytelling for sustainable development in rural communities: An alternative approach. *Sustainable Development*, *28*(6), 1813–1826.

Lowery, B., Vodden, K., & Campus, G. (2019). Analysis of asset mapping initiatives in rural NL. Report submitted to Atlantic Canada Opportunities Agency. Retrieved from http://ruralresilience.ca/final-report_rural-asset-mapping-study-2019/

Mannion, P. (2018). *A land of dreams: Ethnicity, nationalism, and the Irish in Newfoundland, Nova Scotia, and Maine, 1880–1923*. McGill-Queen's Press.

Matthews, R. (1979). The smallwood legacy: The development of underdevelopment in Newfoundland 1949–1972. *Journal of Canadian Studies*, *13*(4), 89–108.

Mayda, C. (2004). Resettlement in Newfoundland: Again. *FOCUS on Geography, 48*(1), 27–27.

Ostrom, E. (2009). A general framework for analyzing sustainability of social-ecological systems. *Science, 325*(5939), 419–422.

Parill, E., White, K., Vodden, K., Walsh, J., & Wood, G. (2014). *Regional asset mapping initiative: Humber-Northern Peninsula-Southern Labrador region*. Grenfell Campus, Memorial University.

Prosper, K., McMillan, L. J., Davis, A. A., & Moffitt, M. (2011). Returning to Netukulimk: Mi'kmaq cultural and spiritual connections with resource stewardship and self-governance. *The International Indigenous Policy Journal, 2*(4), Article 4.

Russell, C. (2022). *Asset-Based Community Development (ABCD): Looking back to look forward* (3rd Edition). eBook Partnership.

Ryser, L., Barrett, J., Markey, S., Halseth, G., & Vodden, K. (2023). Municipal entrepreneurialism: Can it help to mobilize resource-dependent small communities away from path dependency? *Regional Science Policy & Practice, 15*(7), 1477–1493.

Scoones, I. (1998). Sustainable rural livelihoods a framework for analysis. *Analysis, 72*, 1–22.

Slawinski, N., Lowery, B., Seto, A., Stoddart, M., & Vodden, K. (Eds.). (2023). *Revitalizing PLACE through social enterprise* (Issue 41). Memorial University Press.

Smith, M. (1968). *Newfoundland, 1815–1840: A study of a merchantocracy* [Masters, Memorial University of Newfoundland].

Sools, A. M., Tromp, T., & Mooren, J. H. (2015). Mapping letters from the future: Exploring narrative processes of imagining the future. *Journal of Health Psychology, 20*(3), 350–364.

St. Croix, A. (2015). *Final Report on a proposal to provide training and oversight for the implementation of a cultural and heritage resources inventory and website*. Report submitted to the Town of Branch, NL.

Tourism Synergy Ltd., & Broad Reach Strategies. (2016). Gros Morne region strategic tourism plan. The Gros Morne Regional Tourism Working Group. https://grosmornecoop.com/wp-content/uploads/FINAL-GM-STAR-Plan-Nov-23-2016.pdf

Tuck, J. (2022). Beothuk. In the Canadian Encyclopedia. https://www.thecanadianencyclopedia.ca/en/article/beothuk

Upton, L. F. S. (2008). The extermination of the Beothucks of Newfoundland. *Canadian Historical Review, 58*, 133–153.

Van Assche, K., & Gruezmacher, M. (2024). Asset mapping 2.0; contextual, iterative, and virtual mapping for community development. *Community Development, 55*(2), 195–210.

Van Assche, K., Gruezmacher, M., Vodden, K., Gibson, R., & Deacon, L. (2021). Reinvention paths and reinvention paradox: Strategic change in Western Newfoundland communities. *Futures, 128*, 102713.

Van Assche, K., Beunen, R., & Gruezmacher, M. (2024). Strategy for sustainability trransitions: Governance, community and environment. Edward Elgar Publishing.

Van Der Vaart, G., Van Hoven, B., & Huigen, P. P. P. (2018). Creative and arts-based research methods in academic research. Lessons from a Participatory Research Project in the Netherlands. *Forum Qualitative Sozialforschung/Forum: Qualitative Social Research, 19*(2). https://doi.org/10.17169/FQS-19.2.2961

Vodden, K., & Hall, H. (2013). *Understanding regional governance in Newfoundland and Labrador a survey of regional development organizations*. Canadian Rural Revitalization

Foundation, Municipalities Newfoundland and Labrador, Memorial University, University of Kentucky.

Walzer, N., Phillips, R., & Blair, R. (2020). *50 years of community development Vol I: A history of its evolution and application in North America*. Routledge.

Withers, G. (2016). Reconstituting rural communities and economies: The Newfoundland fisheries household resettlement program, 1965–1970. Doctoral (PhD) thesis, Memorial University of Newfoundland.

# 2 Stories and identity in community development

**Abstract**
*This chapter digs deeper into the theme of assets as embedded in community narratives, stories, and identities. It outlines how a shared way of appreciating the world can enable cohesion and coordination of actions but can also be a challenge for diverting the path in a desired new direction. As such, the narrative fabric of a community is understood as an infrastructure enabling the reinterpretation of assets and strengths.*

In discussing the interconnections between stories, narratives, and community identities, this chapter recognizes that communities, however small, have multiple stories and identities despite this heterogeneity often being overlooked (Cresswell, 2004; Daniels et al., 2016). Shared identities can be seen as a set of images and stories that are reproduced by the group, creating a sense of unity but also a distinction from other groups. Stories of identity can create boundaries between "us" and "them," just as physical boundaries can create or reinforce identity. They reflect and are reflected in not only stories and narratives that are told and retold but also in the rules, norms, and procedures that allow communities to coordinate and function as a group.

Identity relies on the memory both of individuals and collectives. The way people remember themselves, their past, pains, and successes, is a defining part of that collective identity (Edkins, 2006). What counts as "history" is then shaped by those social memories and collective identities, as well as narratives both internal and external to a community (Antze & Lambek, 2016; Van Assche et al., 2023). We approach identity narratives both as a source of rigidity and creativity, in which storytelling is an essential element of understanding community past, present, and future. This chapter asserts that rethinking assets cannot be done without the careful examination and often reconsideration of stories, narratives, and identities.

DOI: 10.4324/9781003620297-2
This chapter has been made available under a CC-BY-NC-ND 4.0 license.

## Why do we tell stories?

Stories are a timeless method for recording and transmitting knowledge and have a unique ability to express and ignite the desires of both individuals and collectives. They connect thinking and organizing, as stories can help people see how things could be organized in the present and for the future. A story can persuade more effectively and drive home the importance of something better than facts, partly due to the intimate link between stories and emotions (Harris, 2024), which relies on bringing the listener into the story through a gap in the narrative in which they can project their own experience (Shenk et al., 2023). Stories can portray cause and effect in ways more convincing than formal logic or experiment, and they can connect the small and the large, the local and the global, or the social and the economic or the ecological.

What is understood as an asset for a community comes about through stories of self, environment, and their coupling. The stories at the heart of a community reveal its potential and desires, as well as images of the future, which are fundamental in shaping the way communities organize themselves.

*Figure 2.1* St. John's Storytelling Festival promotes the art and tradition of storytelling in Newfoundland and Labrador, fostering creative expression of this art, and promoting cooperation among organizations related to the art and tradition of storytelling. Poster design Graham Blair.

Co-author and Indigenous scholar Amy Hudson illustrates how stories can express and ignite desires and create futures (Figure 2.1).

*When we tell our own story by reflecting our past from our place and values, we can begin to decolonize ourselves from the stories that have been told about us by outsiders. This has been an important process for me both as a researcher and as a community member. Through this process, I, and I believe that "we," can begin to destabilize the authority that has been held for far too long by the Western European storyteller. For too long, this type of outside actor has told our story (that of NunatuKavut Inuit and our history) and has benefited from the power and authority that comes with its creation (Hudson, 2020).*

Myths are narratives that seek to explain big themes, often related to the natural world or the human condition. Myths can be inspired by or related to political ideologies or cultural values (de Neufville & Barton, 1987). While myths can relay important teachings, sometimes dominant myths can lead communities into blind alleys and dead ends (Žižek, 1989). Myths can become particularly important in challenging environments because they can inspire hope and social cohesion, sometimes paving the way for collective action. Fantasies can play a similar role. If people are unhappy with the current situation, it is easy to imagine alternatives, even when they seem impossible to achieve.

Stories are closely associated with certain types of knowing and organizing the world around us. Privileging narratives about efficiency can inspire minimizing red tape in administration, while narratives that emphasize connecting with nature could incentivize investment in outdoor recreational activities or conservation (Bertelli, 2006). Some stories sharpen observation of the outside world and one's community, while others reduce the need or desire to observe. If we look for inefficiencies, inspired by dominant stories that promote low-cost government, then we will find more inefficiencies, forgetting that the existing form of administration could be useful for different goals and might enable adaptation. Narratives create lenses of observation that reinforce certain ways of knowing and organizing (Czarniawska, 2014; Seidl, 2016).

### Land as strength and the false narrative of the landless band in Ktaqmkuk

Avery Velez

Teluisi Avery, ni'n na Mi'kmaw aqq Portuguese, tleyawi Elmastukwek Ktaqmkuk aqq wetapeksi pkite'snuk Nujio'qonik, pkite'snuk Ktaqmkuk kikjiw Isle aux Morts, aqq Portugal. My name is Avery and I'm mixed Mi'kmaw and Portuguese from what is colonially known as the Bay of Islands in Ktaqmkuk (Newfoundland). My roots are in Bay St.

George South (Robinson's) and the south coast of Ktaqmkuk around the community of Isle aux Morts. I introduce myself in this way as an act of self-location and accountability to nmitkinen aqq msit no'kmaq (our Mi'kmaw territory and all my relations).

Mi'kmaq have been living with the land in Ktaqmkuk according to our own principles and traditions of self-governance since time immemorial. Some Mi'kmaq, the Sa'qewe'jkik (the Ancients), lived in Ktaqmkuk year-round, while others traveled back and forth between Unama'ki (Cape Breton) and Ktaqmkuk. For centuries, British colonial powers and white settler Newfoundlanders have ignored and denied our Indigeneity to the island based on ignorance, racist assumptions, and/or belief that Mi'kmaq are only in Ktaqmkuk because they were brought from Nova Scotia by the French to kill the Beothuk.

Ktaqmkuk has always been, and will always be, part of Mi'kma'ki (Mi'kmaw territory) but Ktaqmkukewaq Mi'kmaq (Newfoundland Mi'kmaq) today have extremely limited recognized rights to live with the land and each other as we have always done. Only Mi'kmaq represented by Miawpukek First Nation have the kind of access to land generally afforded under the *Indian Act*, just over 16 $km^2$ around Conne River. The rest of the Mi'kmaq in Ktaqmkuk are either wholly unrecognized (i.e., Mi'kmaw territory without Indian Status) or registered under Qalipu Mi'kmaq First Nation (Qalipu), considered a "landless" band by the federal government (*Agreement for the Recognition of the Qalipu Mi'kmaq Band*). Yet land is the foundation of who we are as Mi'kmaq and without access to nmitkinen (our territory), our sustainability as a community is under threat.

In 2007, the Federation of Newfoundland Indians signed an Agreement in Principle with the federal government for the creation of Qalipu, a conglomerate *Indian Act* band that subsumed the nine original Mi'kmaw bands existing outside of Miawpukek and was officially established in 2011. Qalipu's recognition was only partial – it allowed certain community members to register as Status Indians but it did not come with any recognition of our relationship with nmitkinen, creating a reputation for Qalipu as a "landless band." Those who signed the *Agreement for the Recognition of the Qalipu Mi'kmaq Band* gave up on land recognition in favor of financial support for health and education. Yet we can learn more from the land than from colonial institutions of education. Further, our community's health and well-being are directly being tied to that of the land we live on. But the Government of Canada held our communities hostage and manipulated our leadership into settling for partial recognition that separates us as a people from what makes us who we are – the land. This landlessness, though written in plain language in the *Agreement*, is a colonial myth. The land is part of us, and as long as we exist and persist, it remains ours to live with and protect, regardless of colonial recognition. As Mi'kmaq, the land

*Figure 2.2* "Mi'kmaq Strength" (2021) by Mi'kmaq artist Marcus Gosse displays a whale (putup) holding up a fishing village. Each of us has the inner spiritual strength to contribute to our culture and community. A whale can communicate with other whales up to 10,000 miles away. May we have the power to communicate with each other this effectively. Let us have the strength to support each other culturally and spiritually. Msit No'kmaq (All My Relations – We Are All Connected).

and our relationships with the land are the core of our culture, nationhood, and our ability to persist into the future (Figure 2.2).

---

When we understand the world through stories we do develop blind spots. If the associated versions of the world are institutionalized, shaping the ideas and policy tools used in governance, these stories illuminate certain paths for the community while obscuring others from view. In the context of fisheries, Kooiman et al. (2005) refer to these as the meta level of governance, where overarching myths and images reside that inform governing institutions and everyday actions. After a while, the way the world appears for the community and for the governance system looks like the only real world and has become naturalized (Van Assche et al., 2024). When stories *about* governance are cemented, the way things have been organized may look like the only way.

## Stories, stability, and change

Connectivity to the world brings in new stories, ideas of the good life, lifestyles, cultural values, and ideologies. In Newfoundland and Labrador (NL), even in communities still without high-speed internet, most people have ample means to learn about the rest of the world. People here have long been connected to the global community through trade, migration, and communications.

Connectivity, and the new ideas and perspectives it brings, can create rifts in communities across many lines, including generational. New stories and desires can also galvanize communities to move in a direction that seems appealing to all. Premier, journalist, and radio personality Joey Smallwood told stories persuading many that life within Canada would open opportunities and enable modern ways of life that extended beyond fishing (Argyle, 2012; Horwood, 1989). However, Smallwood's modernization myth pitted a bright future of prosperity and industrialization against outport fishing life, which was maligned as backward and of the past (Matthews, 1979). This myth directly informed the controversial resettlement program that led to the disappearance of over 150 communities province-wide (Maritime History Archive, 2004). Furthermore, new industries did not always deliver on Smallwood's big promises. Many remained attached to fishing and their outport communities, skeptical of modernization promises, and resisting resettlement.

Stories harden when they are built into decision-making. This can occur because politicians embrace them, or because they have a base of powerful supporters – whether voters, administration, or outside experts – who act as storytellers and reinforce these narratives. They are further strengthened when they become entrenched in administrative structures like departmental mandates, policy topics, and ways of defining problems and solutions. Furthermore, if people are distrustful of planning or long-term strategy, or if institutional structures do not provide adequate capacity, change is harder (Vaara et al., 2016; Van Assche & Verschraegen, 2008). Not only policies and plans but also laws, bylaws, and regulations can enshrine stories as a reality that seemingly cannot be questioned. If environmental quality is understood as only a matter of fining pollution, or spatial planning only as a matter of separating uses by means of zoning bylaws, then a community will struggle to rethink policy goals and assets (Copus et al., 2017; Miller, 2012; Miller et al., 2015).

Stories can be set in stone by other stories (Spicer, 2013). Stories about fishing can be supported by ideas of masculinity, community resilience, and so forth. Decisions to ignore an environmental problem can be based on stories about limited local power and responsibility or fate. Attitudes toward tourism can be grounded in stories about the landscape, cultural identity, or economic development narratives based in external dependence. Ideology and religion often function as embedding stories, upholding or inspiring others in governance, and can inspire belief or rejection of many different topics.

People and communities build a reality – or realities – through stories, which come to support one another mutually to make our world livable and manageable, delineating accepted ways of doing things (Bal & Marx-MacDonald, 2002; Fairclough, 2013). Problems can arise when this seemingly natural state of affairs hides changes in the environment and society or obscures possibilities for the future. Nothing is certain, yet people often create a high degree of certainty around a limited set of stories and observations. As a result, possibilities that may seem obvious to someone not embedded in the fabric of these dominant stories may be unthinkable or missed entirely by those steeped in them.

In framing our understanding of assets, it is important to remember that narratives never stop changing; they are dynamic. Stories change through their retelling, contextual changes, the appearance of new characters or storytellers beyond established roles, or as new connections are formed to other stories and ideas. Thus, *new* stories will always emerge in a community, whether entering from within or outside. New stories can emerge because old ones are forgotten, perhaps no longer retold because their keepers passed away or moved to the margins of society (Antze & Lambek, 2016). Conversely, new developments and changes in the social or ecological environment could be interpreted by turning to the stories of the past.

Stories can also change because they become part of governance, whether simplified by administrative and political procedures or translated into a series of bureaucratic codes (Cohen, 1975; Vaara & Tienar, 2008). When stories are adopted into governance, they enter an arena where competing actors and stories create and are subject to power dynamics that inevitably alter these stories (Miller, 2012). New managers might bring in new approaches, which in turn can introduce new governance stories (Alvesson & Spicer, 2010), creating openings for new forms of participation and topics to enter governance. Similarly, new civil society organizations can both represent and create new stories, distinguishing themselves from other organizations, and possibly introducing these stories into governance (Jessop, 2020).

---

**Canada's only union built town: Port Union, Newfoundland**

Edith Samson

Stories of the past can become valuable for reimagining the future; stories of struggle or conflict can turn into stories of heritage and pride. Unlike other rural Newfoundland outports, Port Union is an intentional community, founded in 1916 by Sir William Ford Coaker and the Fishermen's Protective Union (FPU). It became the headquarters of the FPU and was a self-sufficient community with its own import/export company, hotel, bakery, housing, customs office, post office, newspaper, and woodworking factory until the late 1970s.

In 1992, Port Union faced another challenge with the cod moratorium and the closure of Fishery Products International, the local fish plant. With the displacement of the workforce and outmigration from the area, local groups formed to seek ways to diversify the economy.

Port Union focused on its heritage, considered how it could rebuild, and looked at other economic opportunities such as tourism. The Sir William Ford Coaker Heritage Foundation was founded to oversee the preservation, conservation, and preservation of the Heritage District. The row housing – once fully occupied – and the former home of the Fishermen's Advocate newspaper were closed, and the former salt fish plant and retail store were in a state of disrepair. The hotel and Union Electric office building became part of the Foundation's property. The Foundation began working to restore the former factory building and thus began the transformation of "Canada's only union built town."

With over 30 buildings in the district, it has been a journey of seeking ways to acquire the properties, funding for their restoration, and finding ways to maintain and sustain the heritage properties. Today, the former woodworking factory is the Interpretation Centre for the Port Union National Historic District of Canada. The National Historic District is also a Provincial and Municipal Heritage District.

One of the row house duplex units now houses Union House Arts, a small art gallery that hosts local, national, and international artists. A bed and breakfast now operates in another of the houses by a private individual and leased from the Foundation. The FPU row housing that provided affordable housing to workers is being restored. The local area has a housing shortage, and the Foundation is using its resources to become part of the community's solution.

Working with heritage properties comes with a variety of challenges and successes. There is now less funding available for heritage organizations and fewer donations. However, the vision remains the same. Port Union was based on the idea that fishermen could rise above their dependence on the merchants and seek a better future for themselves and their families. Today, that mission continues through the Sir William Ford Coaker Heritage Foundation's revitalization efforts in the "Union Built Town," with a vision to benefit the local community and create economic opportunities.

## Collaboration and conflict

Conflict marks many compelling stories. Some stories convey aspirations, hopes, and fears, and through their ties to community identities can spark strong emotions. Not all stories in governance touch identities and trigger emotions, and not all aspects of identity narratives trigger sensitivities.

However, this is likely when communities face drastic change and dramatic choices, and where competing stories arise. Few can predict how these emotions might manifest, and which conflicts might be created or awakened (Erikson, 1994; Fink, 1999). Thus, conflicting or changing stories must be treated with care.

Some conflicts are easier to predict than others. When competing ideologies, and often different understandings of key assets and futures, have coexisted for a long time in uneasy tension, and something reinforces either side, tensions can explode. Alternatively, when different stories about the community coexist and this narrative landscape is polarized into two opposing camps, it becomes more difficult to analyze problems in different ways or collaborate in creating shared visions (Cuvelier et al., 2014). Conflict can erupt more easily as the two sides cling to fixed problem definitions and solutions. In these circumstances, coming to new understandings of assets and possible futures can be extremely difficult as identity-based narratives become synonymous with a particular position in a debate and compromise can be seen as betraying that identity (Van Assche et al., 2023).

Not all conflicts are negative. Conflicts can be productive, provided they do not undermine legitimate institutions, inspire violent aggression, or make deliberation and future collaboration impossible. The apparent absence of conflict or tension within a community that is clearly at risk is often a warning sign. Lack of conflict can be the result of an oppressive social structure in which dominant groups have a vested interest in maintaining a *status quo* that disempowers the rest. Shared values and morality might give an impression of stability and individuals might fear destabilizing the *status quo* by raising questions or doubts, thus deterring debate when it is perhaps most needed (Cohen, 1985; Cooke & Kothari, 2001). Aversion to conflict can lead to the suppression of differences and the silencing of critical discussion.

New ideas can emerge in conflict, especially when each party is forced to reflect on their position on the issue and identify creative solutions. This process can be helpful for crafting arguments in support of decisions, including garnering support from external actors like funders. Various opinions can come to the surface, encouraging a more honest sharing of perspectives that existed before the conflict (Hulliung, 2017; Oksala, 2012). Emerging ideas, perspectives, and stories can lead to the identification of new assets and, in turn, contribute to public learning. Conflicts can produce objects that are recognized only in and because of the conflict (Dreu & de Vliert, 1997). A fish might look like any other fish before the conflict but become something else when it is threatened or becomes the focal point of debate.

A different source of instability is shocks. We refer to shocks as radical events that may challenge existing forms of knowing and organizing in the community. Shocks can be *material*, as in an earthquake or the collapse of a fish stock, or *discursive*, as when situations become suddenly impossible to understand or new stories are imposed that do not connect to the current frames of understanding in a community (Van Assche et al., 2022, 2023). Connections between narratives can stabilize or normalize shocks. Uncontested pillars of community identity may be resistant to change. However, structures that are resistant can be brittle, like bridges or buildings crumbling when hit by an earthquake. Shocks or disturbances, whether material or discursive, can transform loosely held narratives as well as those that are highly robust.

Shocks can resonate with community identity in complex patterns. Loosely held narratives do not generally play a strong coordinating role and when met with shock, the void left behind can be filled with other narratives. In other cases, shocks can undermine narratives that are central to governance, social identity, and the relationship between people and the environment. When conflict follows shock, understanding the shock and rebuilding afterward are further complicated (Young, 2010).

In NL, the economic shocks of the early 20th century and then the cod moratorium of 1992 provoked a wide variety of conflicts. At the same time, these shocks sometimes brought communities together in a sense of nostalgia for livelihoods that were perhaps less appreciated when they were dominant (Bavington, 2011; Davis, 2014; Hamilton & Butler, 2001; Rose, 2007). The shock of the quasi-forced relocation of communities in the 1950s and 1960s during the resettlement program has also created its own narratives. For some, narratives of resettlement come in the form of nostalgia for outport life or even anger and trauma about cultural genocide, while others saw this period as a time of welcomed modernization. Relations between rural communities and both provincial and federal governments are still shaped by this experience, often causing mistrust of government institutions and policy initiatives, and limiting the capacity for collaboration that might be necessary to chart a new course (Hoggart, 1979; Loo, 2019; Mayda, 2004). Mistrust linked to past experience is a challenge that extends beyond government relationships to insider–outsider relationships more broadly.

---

**Fallen saviors and newfound strengths: a story about time spent on the Great Northern Peninsula of Newfoundland**

Brennan Lowery

This story is a personal reflection on my time living on the Great Northern Peninsula (GNP) of Newfoundland. It reflects on insider–outsider relations and outsiders seeing themselves as "white knights." I lived in three different communities on and off for about 1.5 years, first while working on my PhD, then afterward as a postdoctoral researcher.

I first went to the GNP when I was invited to go to St. Anthony for a workshop hosted by the Grenfell Office of Engagement and the Harris Centre. In the room, there was frustration over a recently published report on the demographic outlook of the province, and the negative story that was told. Some participants passionately shared all of the good things happening in their communities in addition to the demographic issues. That workshop inspired me to change my PhD research, which I decided to do on the GNP. Over the next year, I made several visits.

I decided to move to the community of Port au Choix. I had changed my focus to start thinking about storytelling and asset mapping as a planning tool. Over two months, I held interviews with about 30 people who shared many stories with me, stories of enthusiasm and pride but also disappointment, especially as they talked about different initiatives over the years that hoped to bring about new economic opportunities to the region. Many projects failed despite promises made by government agencies or business investors. I also heard stories of disappointed hopes in academic research partnerships that ended up being extractivist research that didn't benefit communities, failed to follow up with community partners, or even reinforced deficiencies-based narratives.

Many stories described saviors who promised to alleviate the uncertainty felt by people regarding their community futures – saviors who ultimately fell from grace. At the same time, I began telling myself a story in which I could become a savior of sorts and keep the promises that I made to community members, even if mine were much more modest than opening a new factory or creating hundreds of jobs. I cast myself into a narrative populated by heroes, villains, and victims – vowing to be the researcher who truly engaged in genuine co-creation, who followed up, who earned the trust of community members, and kept it.

I later moved to St. Anthony – the place where my GNP story had begun. It was a time of great hope and promise. It seemed like I finally had the chance to be the hero the GNP needed, and I was showing my commitment in the most tangible way possible. There was just one problem: I wasn't happy. I found myself looking for any opportunity to get away. Both projects that I worked on with community partners failed to live up to the expectations that I and the community partners had for these initiatives. Despite trying to be the hero, and rationalizations that my presence was literally reversing the demographic decline often used to characterize the region, being there was taking a massive toll on my mental health. I packed up my car and moved down the coast to Corner Brook.

Reflecting on that time, I can see many lessons about the stories we tell as researchers about communities "in trouble." More than anything, I learned about the delicate interplay between the assets and needs of the communities we want to help and the assets we possess as individuals. As outsiders in these communities, there are many tensions to navigate regarding our reflexivity and power dynamics, which are well covered in the literature on community-based research. Perhaps a less-discussed tension is the need to be reflexive about the stories we tell ourselves and how they can affect our relationships with community members and our own well-being. I see now that being a savior was never possible. People in rural NL must be their own heroes. Although researchers won't be able to change the story of a community singlehandedly, they can help a community tell its story to audiences that might not otherwise hear it, and maybe along the way be a friend, ally, and partner.

## Insiders and outsiders

When communities feel neglected by centers of power, and livelihoods and social support mechanisms are under pressure, stories about outsiders to the community are also affected. When outsiders show up, whether as new residents, advisors, or government officials, the relationship between locals and these newcomers is colored by the complicated history of past relations (Hicks, 2018). This can often result in distrust. Outsider perspectives may also be difficult to embrace because there are limited avenues for newcomers to get involved in community life, for example, few local organizations in which they can get involved or connect with others.

In rural NL, older residents are often stewards of community governance and dominant narratives. There are many organizations led by older community members with innovative and forward-thinking mindsets. However, it can also be the case that when such stewards are guided by nostalgia for the past and its values and lifestyles, there is little room for, and even resistance to, thinking about different futures. This challenge can be amplified by public discourse or limited governance capacity, with stretched local governance institutions having few tools to envision and act toward change (Parkins & Reed, 2012; Minnes & Vodden, 2017). We will develop this point in the next chapters.

In contexts where community leaders have stopped or find themselves unable to imagine alternative futures, outsiders can bring important new perspectives. Outsiders may see different things in the community and its environment, both assets and challenges. They bring their own stories about quality of life, what makes a good community or an attractive environment, and they are likely to relate to existing local narratives in a different way than long-time residents, tending to interpret and respond to them differently and to connect those stories to others that might not be known to locals (Hiller & Neary, 1980). They might observe how old narratives are built into community life and governance, even stories local residents do not necessarily identify anymore, at least not consciously.

Outsiders may be able to help solve problems, bring new perspectives to entrenched conflicts, or alter the balance of power (Caine et al., 2007; Elias & Scotson, 1994). They can bring useful assets with them, help to replenish a dwindling population, attract investment and expertise, strengthen the labor market, or contribute to cultural richness and diversity. Outsiders can put the community on the map in new ways (Cohen, 1975; Hall & Vodden, 2019; Van Assche et al., 2021).

But newcomer perspectives can just as easily cause damage. There is the well-known "white knight" phenomenon, in which a well-intentioned newcomer can come to see themselves as a savior of a community supposedly lacking something, which can harm both the community and the would-be savior, a phenomenon Brennan Lowery discussed from firsthand experience above. A well-intentioned newcomer can inadvertently aggravate conflicts by aligning themselves with controversial local leaders or taking a side in complex local issues with nuances that they do not fully understand. These harms can multiply when communities are under stress or facing trauma from a history of marginalization by actors outside of the community.

Ultimately, the most important role of outsiders is in observing what is familiar and natural for a community through a different set of eyes, leading to new questions and ideas. Outsiders can sometimes see assets in a community whose value is not recognized or has been forgotten, like treasures hiding in plain sight. Yet, balancing insider and outsider perspectives is far from simple (Devine, 2007; Kerstetter, 2012). At a given time, outsider perspectives might be useful, while at other times they might stoke conflict when the community is not ready or able to productively deal with it. Especially in places where there is collective trauma, outsider intervention in any form, however useful, must be carefully considered and mistrust is to be expected. However, when a community is ready to welcome outside interpretations and ideas, there is great potential for new stories to emerge by bringing together perspectives of both locals and outsiders.

---

**Urban to rural journey**

John Fisher

My family and I moved to Port Rexton 34 years ago. The experience of living here became richer for us when aligning our expectations with the reality of this place. Our relationship with the community is best characterized as well-treated guests. As they say in the outports you can't be bad friends – a simple but profound underpinning for collective living in once isolated communities that faced enormous and sometimes life-threatening odds.

Locals reckoned that no one willingly moves to an outport without a compelling and perhaps a questionable reason to do so. But 30 years ago, a Newfoundland outport that appeared to us to be lost in time was an attractive alternative for those disillusioned with the affluence, excess, cost, and complexities of living the late 20th-century urban dream.

Relocating to another community is relatively straightforward. Moving to another culture – a very different matter altogether. Outports are relatively homogeneous, with strong kinship ties, some still home to ancient dialects and ways of thinking that may appear to the outsider as a nearly impenetrable code. It is a distinct culture.

Outport Newfoundlanders have survived by successfully manipulating the physical environment. In the face of a failed inshore fishery and outmigration, however, the ability to conjure/imagine another future has become another essential asset. With this, communities are left with limited means to sustain themselves. Ironically, those who migrate to these seemingly peaceful places quickly and with the very best of intentions want to get these communities "back on track" – at least economically. It's a fascinating contradiction. We move to live a simpler life but then want to engage with the community using knowledge and skills developed in the place we left behind! Not incidentally, economic development agencies give enthusiastic support to these initiatives led by newly arrived outsiders.

Relocation to anywhere has two parts. The desire to leave and the pull of where we are headed. Fifty-fifty in our case. Our disenchantment with over-development in Ontario and the need for our family to trade disconnected living for a connected community…also the ocean and sky at night. We started a four-room bed and breakfast with a digital shop accessing customers around the world, which quickly morphed into a 32-room inn and conference center employing 35 local people and hosting 12,000 visitors between May and November. The inn is part social enterprise. Our staff are as critical to our success as our guests. We invest in community and support the arts, establishing a unique identity for the inn. We and others founded *Riddle Fence*, the province's award-winning literary journal, and Business and Arts NL.

In Port Rexton, with Town Council support, we initiated and raised funds for a strategic plan for the town. A glimpse into the future seemed useful given colossal changes in the world. Then, it was shelved the way strategic plans are, save the identified need for affordable housing to attract young people. A 23-year-old university educated resident incorporated a not-for-profit affordable housing corporation. The inn donated four acres of land. Over a million dollars has been secured from federal and provincial governments, and construction of a multi-unit accommodation is imminent.

The challenge facing urban people moving into rural places is to replace our utopian assumptions with an appreciation of local views. Perhaps to see ourselves as others see us. Yes, it may be frustrating that our ideas are not embraced by local residents. However, this is an opportunity to start over. Once we figured all this out, the opportunity to live here in peace and part of a community became a gift and privilege that Peggy and I, Luke and Gabe, their partners and children treasure every day.

## Concluding

What a community considers an asset, how it is used, and which policies are based on that asset are intrinsically linked to the stories of the community.

Those stories are often not fully expressed. The way they connect and support each other is often not fully transparent for either community members or outsiders (del Campo & Wali, 2007). Community identities can work as master narratives that produce other stories and give them meaning. Identity can be something of great value, a fabric of stories and embedded values that ties together a community and enable the creation of a shared past and common future (Miller, 2012; Savage, 2008).

Self-organization and strong identities can, however, make it harder to see things differently and adapt to changing circumstances. In this complex landscape of community stories and identities, outsiders wishing to contribute to enhancing local assets must proceed with caution. Outsiders and their perspectives can sometimes be met with mistrust inhibiting their integration into local culture and governance. Strong identities can foster rivalries, even with neighbors with whom collaboration would be useful.

Without a doubt, newcomers to a community can bring a fresh perspective and see things that long-time residents might have taken for granted or missed. In this context, outsiders might offer helpful ways out of gridlock or new ideas for how communities can respond to changing environments or shocks. However, they can just as easily cause harm, even if inadvertently. NL's history of colonial control and core–periphery relationships provides us with a powerful reminder of why mistrust in outsiders exists and is often warranted.

Yet it remains possible for a community to appreciate and cultivate diverse perspectives, to internalize and critically assess outsider perspectives, and to cultivate reflexivity. Fostering places and moments to learn from others' perspectives, perhaps shifting your own, can be difficult but it is not the same as relinquishing or abandoning identities or undermining community values. A narrative that reflects a genuine understanding of local identity, assets, and challenges can both honor the past and speak of alternative futures. The next chapter will consider the role of public discourse, different forms of learning, and education in creating such openings for new perspectives and critical self-reflection for community reinvention.

## References

Alvesson, M., & Spicer, A. (2010). *Metaphors we lead by: Understanding leadership in the real world*. Routledge.

Antze, P., & Lambek, M. (Eds.). (2016). *Tense past*. Routledge.

Argyle, R. (2012). *Joey Smallwood: Schemer and dreamer*. Dundurn.

Bal, M., & Marx-MacDonald, S. (2002). *Travelling concepts in the humanities: A rough guide*. University of Toronto Press.

Bavington, D. (2011). *Managed annihilation: An unnatural history of the Newfoundland Cod collapse*. UBC Press.

Bertelli, A. (2006). The role of political ideology in the structural design of new governance agencies. *Public Administration Review*, *66*(4), 583–595.

Caine, K. J., Salomons, M. J., & Simmons, D. (2007). Partnerships for social change in the Canadian North: Revisiting the insider–outsider dialectic. *Development and Change*, *38*(3), 447–471.

Cohen, A. P. (1975). *The management of myths: The politics of legitimation in a Newfoundland community*. Manchester University Press.

Cohen, A. P. (1985). *Symbolic construction of community*. Routledge.

Cooke, B., & Kothari, U. (2001). *Participation: The new tyranny?* Zed Books.

Copus, C., Roberts, M., & Wall, R. (2017). Policy narratives in local and national government. In *Local government in England* (pp. 37–55). Springer.

Cresswell, T. (2004). *Place: A short introduction*. Blackwell.

Cuvelier, J., Vlassenroot, K., & Olin, N. (2014). Resources, conflict and governance: A critical review. *The Extractive Industries and Society*, *1*(2), 340–350.

Czarniawska, B. (2014). *A theory of organizing*. Edward Elgar Publishing.

Daniels, J., Baldacchino, G., & Vodden, K. (2016). Matters of place: The making of place and identity. In K. Vodden, R. Gibson, & G. Baldacchino (Eds.), *Place peripheral: Place-based development in rural, island, and remote regions* (pp. 23–40). Memorial University Press.

Davis, R. (2014). A Cod forsaken place? Fishing in an altered state in Newfoundland. *Anthropological Quarterly*, *87*(3), 695–726.

de Neufville, J. I., & Barton, S. E. (1987). Myths and the definition of policy problems. *Policy Sciences*, *20*(3), 181–206.

del Campo, H., & Wali, A. (2007). *Applying asset mapping to protected area planning and management in the* Cordillera *Azul*. National Park.

Devine, M. L. (2007). Necessary evils: Strangers, outsiders and outports in Newfoundland drama [Thesis]. https://tspace.library.utoronto.ca/handle/1807/122694

Dreu, C. K. W. D., & de Vliert, E. V. (1997). *Using conflict in organizations*. SAGE.

Edkins, J. (2006). Remembering relationality. In D. Bell (Ed.), *Memory, trauma and world politics: Reflections on the relationship between past and present* (pp. 99–115). Palgrave Macmillan.

Elias, N., & Scotson, J. L. (1994). *The established and the outsiders: A sociological enquiry into community problems*. Sage.

Erikson, K. (1994). *A new species of trouble: Explorations in disaster, trauma, and community*. Norton.

Fairclough, N. (2013). Critical discourse analysis and critical policy studies. *Critical Policy Studies*, *7*(2), 177–197.

Fink, B. (1999). *A clinical introduction to Lacanian psychoanalysis: Theory and technique*. Harvard University Press.

Hall, H. M., & Vodden, K. (2019). Learning, knowledge flows, and innovation in Canadian regions. In Kelly Vodden, David Douglas, Sean Markey, Sarah Minnes, Bill Reimer (Eds.), *The theory, practice and potential of regional development*, (pp. 183–211). Routledge.

Hamilton, L. C., & Butler, M. J. (2001). Outport adaptations: Social indicators through Newfoundland's cod crisis. *Human Ecology Review*, *8*(2), 1–11.

Harris, D. M. (2024). Stories, crisis, and meaning-making: Storying possibility and community in the terrain of cultural struggle. *Cultural Geographies*, *31*(1), 33–46.

Hicks, P. (2018). *Shrinking communities handbook A guide to developing strategies for shrinking communities in Canada*. McGill University.

Hiller, J., & Neary, P. (1980). *Newfoundland in the nineteenth and twentieth centuries: Essays in interpretation*. University of Toronto Press.

Hoggart, K. (1979). Resettlement in Newfoundland. *Geography*, *64*(3), 215–218.

Horwood, H. (1989). *Joey*. Stoddart.

Hudson, A. (2020). *Reclaiming Inuit governance: Planning for a sustainable future in Nunatukavut*, pp. 8–9. [Doctoral dissertation, Memorial University of Newfoundland]. https://research.library.mun.ca/14950/

Hulliung, M. (2017). *Citizen Machiavelli*. Routledge.

Jessop, B. (2020). The governance of complexity and the complexity of governance. In B. Jessop (Ed.), *Putting civil society in its place: Governance, metagovernance and subjectivity* (pp. 35–64). Policy Press.

Kerstetter, K. (2012). Insider, outsider, or somewhere between: The impact of researchers' identities on the community-based research process. *Journal of Rural Social Sciences*, *27*(2), 99–117.

Kooiman, J., Bavinck, M., Jentoft, S., & Pullin, R. (2005). *Fish for life: Interactive governance for fisheries* (J. Kooiman, M. Bavinck, S. Jentoft, & R. Pullin, Eds.). Amsterdam University Press.

Loo, T. (2019). *Moved by the state: Forced relocation and making a good life in postwar Canada*. UBC Press.

Maritime History Archive. (2004). 'No great future': Government sponsored resettlement in Newfoundland and Labrador since Confederation. https://www.mun.ca/mha/resettlement/

Matthews, R. (1979). The smallwood legacy: The development of underdevelopment in Newfoundland 1949–1972. *Journal of Canadian Studies*, *13*(4), 89–108.

Mayda, C. (2004). Resettlement in Newfoundland: Again. *FOCUS on Geography*, *48*(1), 27–32.

Miller, C. A., O'Leary, J., Graffy, E., Stechel, E. B., & Dirks, G. (2015). Narrative futures and the governance of energy transitions. *Futures*, *70*, 65–74.

Miller, H. T. (2012). *Governing narratives: Symbolic politics and policy change*. University of Alabama Press.

Minnes, S., & Vodden, K. (2017). The capacity gap: Understanding impediments to sustainable drinking water systems in rural Newfoundland and Labrador. *Canadian Water Resources Journal*, *42*(2), 163–178.

Oksala, J. (2012). *Foucault, politics, and violence*. Northwestern University Press.

Parkins, J., & Reed, M. (2012). *Social transformation in rural Canada: New insights into community, cultures, and collective action*. UBC Press.

Qalipu Mi'kmaq First Nation Act, S.C. 2014, c. 18. https://laws-lois.justice.gc.ca/eng/acts/Q-0.5/page-1.html?wbdisable=true

Rose, G. A. (2007). *Cod: The ecological history of the North Atlantic fisheries*. Breakwater Books.

Savage, M. (2008). Histories, belongings, communities. *International Journal of Social Research Methodology*, *11*(2), 151–162.

Seidl, D. (2016). *Organisational identity and self-transformation: An autopoietic perspective*. Routledge.

Shenk, L., Franz, K. J., & Gutowski, W. J. (2023). Minding the gaps. *Environmental Humanities*, *15*(3), 83–103.

Spicer, M. W. (2013). Contesting narratives in politics. *Public Administration Review*, *73*(5), 768–771.

Vaara, E., Sonenshein, S., & Boje, D. (2016). Narratives as sources of stability and change in organizations: Approaches and directions for future research. *Academy of Management Annals*, *10*(1), 495–560.

Vaara, E., & Tienar, J. (2008). A discursive perspective on legitimation strategies in multinational corporations. *Academy of Management Review*, *33*(4), 985–993.

Van Assche, K., Beunen, R., & Gruezmacher, M. (2024). *Strategy for sustainability transitions: Governance, community and environment*. Edward Elgar Publishing.

Van Assche, K., Gruezmacher, M., & Beunen, R. (2022). Shock and conflict in social-ecological systems: Implications for environmental governance. *Sustainability*, *14*(2), 610.

Van Assche, K., Gruezmacher, M., Marais, L., & Perez-Sindin, X. (2023). *Resource communities: Past legacies and future pathways*. Taylor & Francis.

Van Assche, K., Gruezmacher, M., Vodden, K., Gibson, R., & Deacon, L. (2021). Reinvention paths and reinvention paradox: Strategic change in Western Newfoundland communities. *Futures*, *128*, 102713–102713.

Van Assche, K., & Verschraegen, G. (2008). The limits of planning: Niklas Luhmann's systems theory and the analysis of planning and planning ambitions. *Planning Theory*, *7*(3), 263–283.

Young, O. R. (2010). Institutional dynamics: Resilience, vulnerability and adaptation in environmental and resource regimes. *Global Environmental Change*, *20*(3), 378–385.

Žižek, S. (1989). *The sublime object of ideology*. Verso.

# 3 Public discourse and social learning in the rethinking of community

**Abstract**

*As the previous chapter illustrated, rethinking assets often entails reinterpreting old narratives, or telling new ones, about the community's past, future, and environment. This becomes challenging when dominant stories presume that change is unnecessary or undesirable. Similarly, when local traditions and underlying narratives label speaking out in public or even having an opinion about community futures as inappropriate, capacity for collective strategizing is greatly limited. When the understanding of community goes unchallenged, the possibilities for learning and adapting are limited. Local understandings of good governance that place the future of the community in someone else's hands are also problematic. Strong local governance and institutions can aid in processes of reimagining community futures and creating paths toward them. This requires an openness to learning and adaptation as key elements of governance, not only valuing individual achievement and formal education but also learning as a community and in governance.*

Master narratives and norms of the community are often implicit and not discussed. They exist in the realm of the collective unconscious. Hidden narratives are difficult to bring into the open; they tend to resist being articulated in public discussion and do not open to questioning easily (Fink, 2013). Nevertheless, examining these stories can open the community to forms of learning and change that are not imposed from elsewhere but arise organically. *Learning*, in other words, is something that must be understood and fostered if a community wants to step back and re-evaluate itself and its assets (Assmann & Czaplicka, 1995; Boissevain, 1966; Cohen, 1975; Hillier & Gunder, 2005; Yanow, 1992).

We distinguish here between *individual learning*, *social learning* (learning as a community), and *governance learning* (learning processes taking place in

DOI: 10.4324/9781003620297-3
This chapter has been made available under a CC-BY-NC-ND 4.0 license.

the governance system itself). Education, whether formal, informal, or through experience, is of course an important site for learning. Yet, learning takes place in many ways, from simple observation to imitation (consciously or unconsciously) and avoidance of what is threatening or risky. Listening, telling, and retelling stories are ways of learning, as are daily informal interactions where community values are both enacted and learned (Bull et al., 2008; Scott, 1968).

This chapter discusses learning processes in communities, especially in governance, that play a role in rethinking assets. Governance is important, because it is here that some stories and understandings become codified, accepted, and translated into collective decisions, policies, and institutions. We discuss public discourse as it supports and feeds into social learning and learning in governance. Public discourse and learning, therefore, must be considered a distinct type of infrastructure that connects the world of stories and the domain of governance.

### Gros Morne National Park, The Most Beautiful Place: Learning and public discourse crossroads and natural assets

Michael Burzynski and Anne Marceau

Anne and I moved to Rocky Harbour, a small enclave community in the center of Gros Morne National Park, on the west coast of the Island of Newfoundland, in 1989. We came to work with the park in the interpretation and science sections, and in 2012, we retired here since the area had the perfect combination of friends, nature, community, and lifestyle.

The park was established in 1973 – only 16 years old when we arrived – and the information available about it was mostly in technical reports. While working for the park, I read my way through the historical, biological, and geological reports in the park library. Anne and I spent much of our free time hiking and photographing, and we sought out every visiting biologist, geologist, geomorphologist, archaeologist, and historian that we could meet. We made notes during field trips and talks with those experts, and we also maintained a list of questions frequently asked by visitors and tour guides.

The guidebook, *Gros Morne National Park, The Most Beautiful Place* (Burzynski, 2017), began as my external memory – a sort of encyclopedia of park information in which I stored the facts and stories that would have helped us understand the park when we first arrived. Rarely do I leave home without a camera. I use photography as a way to slow down and carefully observe the world around me and record what I see. Over the years, I amassed a large collection of photographs of the area – everything from close-ups to landscape panoramas, and from underwater to aerial views, so I was able to include over 1,000 photographs to illustrate the park's landscapes, geology, animals, plants, and weather phenomena.

*Figure 3.1* Illustration by Agnes Marion Ayre (1890–1940) local flora of Newfoundland. Ayre, A.M. (20th century) 0510 FM Artwork meadowsweet Herbarium 21017 Ayre. Agnes Marion Ayre Herbarium Artwork Collection (36.jp2).

To my mind, one of the most important parts of the book is the Acknowledgments section. I am grateful to the many people with whom we explored the park over the years, the subject specialists who shared their expertise, the visiting writers, musicians, and other artists through whose eyes we have seen this place in a different light, and the many community residents who shared their stories and memories. The Acknowledgments show that the guidebook is not mine, but it is ours. I hope that the guidebook helps residents to understand and share the landscape around them and encourages them to look more closely at the natural world and to continue to be the guides and stewards of this beautiful place (Figure 3.1).

## Public discourse

Rethinking assets and narratives becomes more tangible and meaningful when the process is visible and accepted, when it is supported by public discourse and partly plays out in it. Public discourse can emerge in and through civil society organizations, religious organizations, youth clubs, pubs, stores, and playgrounds. Protest can be a form of public discourse, which can in turn influence governance discussions. Yet if changes in governance and stories only come about through protest or changes in formal politics, the options for adaptation are limited.

Public discourse is always present in some form or another, but in some places, it remains restricted to the reproduction of dominant stories promoting the values and positions of authority and elite. Elsewhere, public discourse elicits ideas, assesses them, and generates new ones, then informing governance – or not (Dunlop & Radaelli, 2020; Yanow, 1992). Even in a democratic society, governance can be relatively closed to public discourse, as when definitions of legitimate public discourse are restrictive, and pressures may make people think twice before speaking out.

When does discourse become public discourse? Whenever something is heard by others and starts to circulate in the community, this functions as *de facto* public discourse, even when the views are not intended to be public (Crow & Lawlor, 2016; Wash, 2020). Intentional public discourse involves an intention for something to become known to a larger audience or the community at large. Unintentional public discourse can destabilize governance and therefore can be unhelpful to public deliberation and governance. This happens, for example, when what was intended to be private becomes public, which can then undermine trust in the spaces of intimacy in social life. On the other hand, when certain information previously unknown to the public become part of public discourse, perhaps shedding a more realistic light on key figures and their views, this can be useful. If people in a community acknowledge a need to get together and discuss public affairs beyond formal political processes, conflict might come to the surface but also a shared need to address tensions and find common ground may be addressed (Abelson et al., 2003; Hulliung, 2017).

There are many ways in which discourse can become public, from grocery store chats to conflicts in a parent–teacher association. Gossip can reveal what otherwise cannot be exposed. Anthropologist James Scott called gossip a weapon of the weak in cases where those in power do not leave many openings for critique (Scott, 1985). Public meeting spaces can be places for the formation of counter-discourse, stories against the ones promoted by elites, even when not promoting a particular view. Schools can contribute to public discourse simply by bringing young people together and exposing them to other ideas, places, and traditions, which can render perceived truths a bit less persuasive (Bellino & Williams, 2017).

Discourse can become public when it is picked up by media and especially when those media show a real interest in community conversations. Even so, unintentional media contributions to public discourse occur as well – when not journalists but readers give something a new relevance (Howley, 2009). In Newfoundland and Labrador, even in small and remote communities, local media has been around for centuries. Local newspapers existed since the 18th century, expressing local opinion, but also confronting it with events far away and alternative values and ideologies, keeping the community embedded in international networks. Regional and national newspapers, radio, and TV stations all became commonly accessible across the province (Whelan, 2002). In recent decades, local newspapers have shut down or been purchased by larger media conglomerates, as well as losing ground to social media, destabilizing their role as a forum for venting and discussing opinions about local matters (Lankshear & Knobel, 2011).

In small communities, discourse tends to become public more easily, picked up by those involved in local governance, both formally and informally. In contrast to the national level, where it is possible to pander entirely to a base of ideological support, in small communities, people tend to understand that some form of mutual understanding and cooperation is needed to keep the community afloat. Small community life also makes it harder to avoid people with opposing views or allegiances.

Moreover, people tend to have different roles at the same time in small towns and villages. In rural Newfoundland and Labrador community leaders often "wear many hats." Close social bonds can be extremely helpful in difficult times, as they can reduce conflict and cement collaboration. They also assist in making discourse public and enabling social learning. At the same time, dissent and difference may become harder to express. When dissonant feelings are harbored, they find few places to surface, particularly when rigid expectations prevail. Public discourse and social learning, in other words, are both supported and limited in tight-knit communities (Van Assche et al., 2014).

Opportunities to bring out alternative views might therefore need to be created purposefully. Dire circumstances might convince those in power, or rival factions, that special public events ought to be organized, or that special forms of self-analysis should be undertaken. When discussions go beyond venting pre-existing opinions and allow for sharing new perspectives, public discourse can enhance social learning. Furthermore, if events or self-reflection are linked to governance, they can forge a link between public discourse and governance – between social learning and governance learning. Thus, public discourse can be more than the sum of pre-existing voices (Fischer, 2003; Pahl-Wostl et al., 2008).

If political and cultural polarization exists, politics may impede the collective examination of assets and futures. When public discourse exists only as conflict between individual interests and governance plays to the middle to appease the masses, then public discourse is not realizing its full potential for learning. Under such circumstances, forging a cohesive and supported community vision that recognizes assets and strengths in a new way will be highly unlikely (Carpentier, 2011; Crow & Lawlor, 2016). In contrast, where an iron-clad consensus exists, perhaps tied to a rigid identity narrative, well-functioning local media and public discourse can help a community question these rigid assumptions. Learning about other perspectives in the community

EL GRECO
From East to West Our Homemade Pizza is the Best
896-8060 or 896-3473
We Deliver

LABRADOR'S REGIONAL NEWSPAPER
THE NORTHERN REPORTER
established in 19[illegible]

DEC 10 1992
PERIODICALS

# TLH - ROMPKEY

With the House of Commons now back in session, Labrador M.P. Bill Rompkey has submitted a private member's motion calling on the Government to designate the Trans - Labrador Highway as part of the National Highway Network and to expedite its completion from Labrador West to the Labrador Straits.

Rompkey has ben urging the Federal Government to put into place a national public works program, which would include the Trans - Labrador Highway, to help pull the country out of the recession and earlier this month he wrote John Crosbie, Newfoundland's representative in the Federal Cabinet, asking that funding for the Trans - Labrador Highway be part of any forthcoming national program for highway construction. It has been Rompkey's contention that federal funding was needed for the Trans - Labrador Highway project: "Federal money started it; federal money must finish it."

"From both economic and social points of views I believe it is high time that Labradorians had a highway that meets national standards linking them to the rest of the country," said Rompkey. "And furthermore, no highway system can truly be considered 'national' until it includes the Trans - Labrador Highway. I am, therefore, asking the Federal Government to make the designation of the TLH as part of the National Highway System a priority."

On Wednesday of last week, Rompkey expressed his pleasure with the Federal Government's announcement of funds for improvements to the TLH. "While it has taken some considerable length of time, I am glad that the Federal Government has finally demonstrated a commitment to upgrading and completing the road system in Labrador and has included it with other priority infrastructure projects to be undertaken across the country."

"I will be asking questions over the next few days," indicated Rompkey, "to determine how much funding we can expect for the TLH and how soon we can expect it." Only $2-billion has been set aside for a long list of infrastructure improvements, including the Trans - Labrador Highway, to be spent over a five - year period. "We don't know how much we'll get," said Rompkey, "but that they have recognized it as a Federal responsibility is good news."

The Labrador M.P. believes that the Minister's statement, if followed with direct action, is a logical next step in the development of the highway which was started when Rompkey was a member of the Federal Cabinet. "I have always advocated the completion of the Trans - Labrador Highway from Labrador West to L'Anse au Clair. We have had the support of Premier Wells and he has communicated his government's support directly to the Prime Minister. Now that we have this commitment in the mini - budget, hopefully funding will be made available so that work can begin as soon as is practical."

## RCMP REQUEST ASSISTANCE

The Royal Canadian Mounted Police are requesting assistance from the person who witnessed the accident and took a licence number of a vehicle that struck another vehicle in the Northern Store parking lot on November 13,1992. Would you please contact the R.C.M.P. at 896 - 3383, and refer to our file no. 92 - 3473. Your assistance in this matter would be greatly appreciated.

# INNU AND INUIT SHOW PROGRESS

(Nain) The Labrador Inuit Association and the Innu Nation announced today that the overlap negotiations between the two nations were making very good progress. The latest negotiating session was held in Nain, from November 25 to November 27.

"Negotiating with the Innu is a relief from negotiating with the governments. There is no need to become obsessed with process like we have to do with governments. We have made progress on the substantive issues from the very beginning" said Toby Anderson, Chief Negotiator for the LIA.

"The tone of these negotiations is very different from those with governments of Canada and Newfoundland. The good will and honest exchanges, and our common aboriginal understanding allows us to make good progress. There is trust and mutual respect, and neither party is trying to take advantage of the other, or to wear them down by unending technical arguments and posturing" said Daniel Ashini, Director of Innu Rights and Environment and Chief Negotiator for the Innu team.

Both parties expressed concern about the unwillingness of the governments of Canada and Newfoundland to resume negotiations with the LIA and Innu Nation. Ashini and Anderson jointly declared, "We are proceeding with these overlap discussions, and this will help minimize the delay caused by the failure of the governments of Canada and Newfoundland to come to an agreement. Unlike the negotiations with Canada and Newfoundland which have dragged on interminably, the overlap negotiations are going well, and we are concerned that the continued [illegible] and unwillingness to take action by these governments will affect our people and our rights."

Canada suspended negotiations both with the LIA and the Innu Nation this May, citing the unwillingness of the Newfoundland government to come to any accommodation on a bilateral agreement on the financial contribution that each government would make to final land claims settlements. Neither the LIA nor the Innu Nation accepted Newfoundland's proposal to continue with their respective negotiations without Canada at the table. They have decided, instead, to concentrate on LIA/Innu Nation overlap negotiations for the present.

The overlap negotiations [illegible] the LIA and the Innu Nation will help clarify the boundaries of the respective territories, identify shared areas of land, set out the rights and responsibilities of the two parties, and clarify how the parties will deal with third parties who wish to proceed with developments in the Innu and Inuit territories. This is the third session of the overlap negotiations held between LIA and the Innu Nation. At this session, the two nations discussed principles upon which an overall agreement would be based.

New Schedule Goose Bay to St. John's
Depart Goose Bay 6:40a.m.
Depart St. John's 5:10p.m.
Most days
usual restrictions apply
Contact Air Nova or your travel agent for details
AirNova
The choice of frequent flyers

HARVEY'S TRAVEL
Glenn Plaza, Happy Valley 896-3389
1-800-563-3496
Open Monday to Friday 9:00 to 5:00
By appointment only —
Thurs. 6:00 to 8:30

TERRINGTON CO-OP
CO-OP
Servicewith a smile"

*Figure 3.2* Last issue of the Northern Reporter, a local newspaper in Labrador. The Northern Reporter (1992-12-07). The Northern Reporter. [text/newspaper]. The Northern Reporter (https://dai.mun.ca/PDFs/northernrep/NorthernReporter19921207.pdf).

can aid the process of rethinking while also enhancing cohesion, enabling communities to look forward as a collective (Jeppesen, 2016) (Figure 3.2).

## Fragmentation and blindness

Whereas diversity of voices in public discourse is generally beneficial, fragmentation of public discourse is usually harmful for collective visioning or community development (Escobar, 1984). As media sources proliferate, their audiences tend to splinter. Media may identify with one ideology or with nothing at all (except perhaps a profit motive) and tend to repeat talking points or play to base interests. In this scenario, vilification of opposing views is commonplace, which discourages genuine public discourse in favor of polarization (Newig et al., 2016).

Polarized communities struggle to come to a shared understanding of important assets for the future. Digital and social media and their algorithms tend to amplify biases. Public discussion relocated to cyberspace has great difficulties in finding its way to actual public spaces and performing the more constructive functions of public discourse. Maintaining healthy spaces for public discussion and deliberation can translate into healthy governance systems (Carpentier, 2011; Newig et al., 2016).

Quality public discourse can foster good governance via social learning, as a key function of governance is to bring out different perspectives, make them visible to each other, and allow for comparison and learning (Bull et al., 2008; Dunlop & Radaelli, 2020). Public discourse might overlook problems, opportunities or uncomfortable truths the community does not want to be confronted with (Dunlop, 2020). However, if platforms are carefully curated, it can also do the opposite, confronting the community with itself, first by showing unknown or disavowed points of view, next by facilitating discussion on them.

---

### Journalism, local media, and community in Newfoundland and Labrador: The loss of healthy public discourse

Sue Newhook

When Newfoundland joined Canada in 1949, the Canadian Broadcasting Corporation (CBC) replaced the Broadcasting Corporation of Newfoundland. Historian Jeff Webb quotes future premier Joey Smallwood's assurances to fellow delegates at the National Convention:

> ...in so far as being a Newfoundland station ...our (BCN/CBC) station would go right on as though confederation had never come at all. There is no intention of trying to Canadianise us...We will have our own programmes, our own local news, our own Newfoundland broadcasting.
>
> (Webb, 2008, p. 190)

And for many years, we did. I grew up in the 1970s and 1980s watching the province's attachment to the CBC. As a teen in St. John's, I was often introduced as "Shirley Newhook's daughter"; my mother's popular community-interview show, Coffee Break, was part of a web of CBC programs that connected with communities around the province. Later I became a journalist myself, working in radio current affairs and at the supper-hour newscast Here and Now. In the 1980s, Here and Now was the CBC's hugely successful provincial flagship, but it was also part of an armada, along with Newfoundland and Labrador's six radio stations across the province and more than a dozen news, current affairs, and entertainment shows on TV.

One element of their success was travel budgets, which allowed radio and TV crews to visit the farthest-flung outports. On road trips to communities with no hotels (in a time of few B&Bs), we bunked in with local families. Eating at viewers' tables, visiting their fish plant, made the connection between the CBC and its audiences direct and personal. The production studio in St. John's reinforced the promise of "our own Newfoundland broadcasting" with arts and entertainment programs that helped power the explosion of regional culture called the Newfoundland Renaissance. People saw and heard their everyday lives on the air, every day.

Calls from viewers commenting on or pitching stories were constant. Journalist and host Anne Budgell remembers callers asking the CBC to take up everything from fish plant quotas to housing problems. Satisfied callers made for loyal audiences – people who would return a reporter's call or help her out on a story. Often, I'd hear, "Would you be Shirley's daughter?" It was a kind of reference check: they might not have met me, but if they didn't know Mom, they often knew someone who did; a neighbor's niece had been on her show and liked her, so I was ok too. Most of my colleagues had similar webs of first-, second-, and third-hand connections with viewers.

Here and Now wasn't the only source of solid journalism at the CBC. But as the flagship, it drew attention and bemusement from some network managers: how could such a large percentage of the whole province be watching one show every night? Maybe a lack of competition, they sniffed. My sense then and now is that they either were unaware or discounted the value not just of the journalism, but of connections that were unimaginable in central Canada's bigger cities and more transient newsrooms.

In early December of 1990, it all changed in an hour. CBC president Gérard Veilleux announced the biggest cuts in the corporation's history – among them, the shutdown of all regional TV programming except news. The cuts were immediate – not even a final show to say goodbye. There were petitions, public demonstrations, and a few small rollbacks, but the old relationship was fractured.

In time, it became one sharp shock among many for journalism and broadcasting everywhere. As digital communications have exploded, the cost of in-person connection has made it a luxury for local and regional news outlets decimated by changing markets. As Ashley Fitzpatrick emphasized: "You know what I don't trust? I don't trust a news outlet who's telling me what's happening in my back yard (while reporting) out of Halifax" (Harris Centre, 2024, 32:33).

A challenge that young journalists and all of us face is to find ways to revive and maintain those connections. There are encouraging examples and experiments; some will survive and others won't. The job is far harder than it was 40 years ago, but the presence of "our own programmes, our own local news, our own Newfoundland broadcasting" and media of various kinds is key to maintaining a thriving sense of community in Newfoundland and Labrador.

## Learning in governance

Governance systems, just like individuals, can learn intentionally, meaning they can actively and deliberately observe, use, and develop knowledge, search for and analyze problems and opportunities. Learning can also take place unintentionally, when through the functioning of the system, the following of its own routines, it changes and adapts to new circumstances (Armitage, 2005; May, 1992).

In this chapter, we focus on situations where unintentional learning and routine forms of intentional learning together are insufficient for prompting a change in governance. The judgment that new forms of learning are needed cannot be made by outsiders, as this would not likely be locally accepted or connect to local knowledge, narratives, and identities, and would not likely fit into existing ways of changing governance. Local recognition of the need for governance learning and change, through public discourse and leadership, is essential (Gray & Sinclair, 2005). Whatever that helps a community to see different forms of learning in governance and community is beneficial for a local assessment of this infrastructure for the rethinking of assets. If a community can discern for itself the functioning of public discourse, and the way it supports learning or not, then a truly significant step has been made toward the rethinking of assets.

For small communities facing big challenges, with small budgets and administrations and limited access to external expertise, *comparative learning* comes naturally, as most often people are aware of other places with similar assets and challenges. In Newfoundland and Labrador, the cod moratorium and the similar starting conditions of many communities created a set of similar problems, and people were aware of them. With the integration of Newfoundland and Labrador into Canada, features of governance models found in the rest of the country were adopted. Provincial elections created networks between politicians and a provincial lobby system, meaning that more possibilities for comparison became possible (Bannister, 2003; Crosbie, 1956).

Learning from others comes naturally by simply observing others. It is an elementary way of assessing our own condition as an individual and as a community. The more we know those that we observe, the better we can understand them, the better the quality of the comparison and more helpful in seeing what lessons may or may not be helpful for our own notions of quality of life. Learning from other places requires a realistic self-understanding to see which external practices or policies might fit one's context. Simply copying methods or practices of governance, including methods to strategize for the future or to map and build on assets, can be counterproductive.

---

### Newfoundland and Norway: learning and networks

Kristof Van Assche and Robert Greenwood

After Newfoundland & Labrador became a Canadian province, a flood of studies appeared, first under the auspices of the governments in Ottawa and St John's, later often by Memorial University researchers. Questions of modernization, of "development," but also of heritage, culture and identity, and the assets and potential assets of the new province were carefully considered. In this effort, comparative learning was important, and Norway was seen as a commendable case.

Finn Frost was a Norwegian forestry scientist from the Agricultural University of Norway, with connections to the émigré community in the US, experience in Scotland, and considerable experience at the Norwegian national forestry agency. Then he moved to Newfoundland, where he worked for government and lectured at Memorial University. He wrote about forestry in Newfoundland, agreeing with the commission on forestry that the sector had a future, that the competitive advantages outweighed the rather slow growth of trees (Frost, 1958).

Aasulv Løddesøl worked on the bogs and peatlands of Norway in the 1940s, and in the 1950s acted as an advisor for the provincial government. Løddesøl recommended that Newfoundland could drain many bogs and make them productive for forestry or agriculture, especially fodder. He recommended an experimental farm on peatland, which the provincial government established (Løddesøl, 1955).

Cato Wadel, a Norwegian anthropologist, packed for Memorial University in the 1960s and became well known for his analysis of work, employment, and unemployment. Drawing from ideas rooted in his Newfoundland experience, and back in Norway, he wrote a book on the psychological and social effects of unemployment and compared the fisheries of Norway and Newfoundland, including their economic organization and policy frames (Wadel, 1968). Wadel was brought in by anthropologist Robert Paine, a professor at Bergen University before coming to Newfoundland. A keen student of Arctic cultures and the processes of modernization and disruption affecting them, he studies traditional cultures in Norway (often Saami culture) and in the Canadian Arctic along with modernization in the province of Newfoundland and Labrador.

Ottar Brox was also encouraged by Paine to come to Newfoundland, as visiting professor. He produced a book on the harsh distinction between modernizing fish processing and traditional fisheries in the province, as well as notes on the Smallwood resettlement program. For Brox, dispersed settlement was not objectively a problem to be solved, but a problem constructed in the rhetoric of politicians with an interest in reshaping social identities.

In the 1980s, Robert Greenwood, later director of the Harris Centre at Memorial University (a successor of ISER), conducted his doctoral research comparing Northern Norway and Newfoundland and Labrador in terms of social, economic, and political organization and the combined effects on economic development (Greenwood, 1991). Svein Jentoft, fisheries social scientist at Tromso University, was a key contact and remained in close collaboration with scholars at Memorial University, notably geographer Ratana Chuenpagdee, analyzing and documenting the enduring relevance of small-scale fisheries.

What started as a rethinking of the economic future of Newfoundland and Labrador in the 1940s and 1950s, and capacity building in the new province, led to networks that enabled comparative research and learning with Norway. Researchers' observations focused on traditional culture, transformation, and the impact of existing policies, sometimes critical, sometimes with an eye on policy recommendations that drew lessons for policy formation from the Norwegian experience.

A second form of learning, which can be inspired by the first one, is *reflexive learning*, or learning from one's own past and experiences. This might be the most demanding form of learning, as it implies an ongoing evaluation of processes and outcomes, and of self (Newig et al., 2016). Reflection is difficult as it means looking at oneself as objectively as possible, even when tempers flare or conditions are tough. It is also difficult because reflections on a community self and its past are unavoidably colored through the lenses of local values, stories, and dreams. Not all governance systems are equipped to critically reflect on themselves, their practices, and decisions. Maintaining diverse sites of observation and self-observation can help, implying the need for a place for community discussions about governance and learning for and through governance (Gunder, 2005; Ostrom, 2005).

Outsiders can provide a different perspective and kick-start critical self-reflection. They may be administrative experts, academics, or consultants, or people from other communities. People from other places can understand a community from outside the stories that frame local residents' experiences. Local knowledge, often forgotten in old-fashioned central planning, becomes even more valuable if it can question itself, and in some cases non-locals are needed to help facilitate that. Bringing in outsiders comes with risks and benefits, and taking any expert opinion at face value can be as damaging to self-reflection as closing the doors completely to outsiders (Elias & Scotson, 1994; Stoddart & Graham, 2016). For comparative learning, reflexive learning, and outsider learning alike, *connectivity* makes a difference. In Newfoundland and Labrador, the hundreds of outports dotting the coasts were never entirely isolated, but the proliferation of media since the early 20th century, adoption of Canadian political and administrative structures since 1949, growth of the public education system, and the construction of physical infrastructure like roads and utilities, rapidly intensified the movement of people, goods, and ideas (Cohen, 1975; Sider, 2003; Webb, 2008).

Learning through *experimentation* is yet another way in which governance systems can learn. In governance systems where organizational capacity, time, space, or resources are limited, experimentation can help communities try out a new approach, policy tool, method, or perspective (Brouwer & Huitema, 2017; Nair & Howlett, 2017). A community is never a laboratory even if it is called a "living lab." One can never create laboratory conditions, keeping all other variables constant and measuring success objectively. There is always a risk with experimentation, which is by definition something new so effects, at least are hard to predict. In a community context, governance experiments can pertain to processes of coordinating people, resources, and forms of knowledge. On the other hand, using the label "experiment" can reduce risk, in rhetoric and in practice. Limiting the scope of the set-up, creating a "test site," a "pilot project," or allowing the whole community to try something which could be a pilot for larger policy reforms, can be attractive both locally and for higher-level governments. The implications and investment look more manageable, even if they appear more unpredictable at the same time.

Turning back to the topic of public discourse, a final form of governance learning to consider is *dialectical learning*. Often taken for granted, dialectical learning, or learning through discussion and deliberation, is an essential feature of governance. It is a naturally occurring form of learning where opposing perspectives have the chance to substantiate and transform their views through well informed arguments, something expected in democratic systems (Ansell & Torfing, 2021; Held, 1996). Although dialectical learning requires differing sides to bring up coherent arguments, it is, of course, not separated from emotion.

Small communities with dense social networks and strong identities have both limitations and benefits when it comes to governance learning. While learning may be generated and shared easily in close-knit communities, fewer resources may be available to institutionalize a variety of forms of knowledge and modes of learning in governance. Differentiating roles and responsibilities in ways that contribute to a diversity of supported and distinct learning processes is difficult without resources that come with size.

Each governance system is marked by different learning forms and sites. Traditions of administration and decision-making privilege some forms of learning and not others (Held, 1996). Management of knowledge flows and translation into policy will diverge. Well established forms of knowledge and narratives will determine the forms of learning that are appreciated, believed, and accepted in a community.

Although we often ascribe a positive connotation to learning, it is important to recognize that governance systems and individuals can also learn to play the game of governance and manipulate public discourse for private or factional benefit. Such learning that deliberately ignores the public interest has been labeled as *dark learning* (Dunlop & Radaelli, 2020; Howlett, 2020; Van Assche et al., 2021). All forms of learning can be abused, and they often combine with what we call *non-learning*. This is not simply ignorance, but a willful refusal to understand and learn because it is convenient, helpful for a career or for one's group or interests. Dark learning and non-learning can obscure the common good, rendering governance inefficient, ineffective, and/or undemocratic (Acemoğlu & Robinson, 2016).

Small towns facing big challenges, with limited resources to mobilize, once again are both at an advantage and a disadvantage. On the one hand, checks and balances can be weaker, along with public scrutiny, if certain shared narratives, values, and futures are unquestioned or certain individuals and groups are undisputed (Brenneis, 1988; Van Assche & Lo, 2011). On the other hand, as more people see more of each other in a small and interdependent population, few things elude collective observation. If dark learning and non-learning go beyond what is truly valued as communal, a small town can turn swiftly against self-serving leadership that routinely relies on not seeing, thinking, or learning. Misguided policies are often harder to ignore in a small town reliant on its natural environment and social capital.

**Extension services and the importance of maintaining different forms of social learning**

Fred Campbell

Memorial University of Newfoundland (MUN) was upgraded to a university in 1949 by Premier Joey Smallwood. He was inspired by the work of the Department of Extension at St Francis Xavier University in Antigonish, Nova Scotia and looked to establish "…an Extension Department with a University tacked on to it, not the other way around" (Jamieson, 2003). Through the Carnegie Corporation, an allotment of $5,000 allowed for traveling libraries to distribute books to 150 communities from the southern Avalon to Cartwright, Labrador through a coastal boat. This was the beginning of the Extension Service concept.

One could say that MUN Extension Services finally got off the ground when Donald Snowden, with experience facilitating co-operatives among the Inuit for the Cooperative Union of Canada, came to Newfoundland and then became director of Extension in 1966. Snowden and his colleague, Tony Williamson, perceived Extension field workers as practitioners of community development and education. They believed that Extension workers must have flexibility in their work, focusing on adult education and processes that enable the individuals of the community to develop, to choose their own priorities and tasks, to be self-determined. The aim was not to tell people what they should be doing but rather listen to their needs and plan with them, not for them.

Extension's last brochure (1989) identified eight field offices across the island with the mission of helping the people of Newfoundland and Labrador realize their potential by engaging with them in a process of social animation and by providing learning opportunities. In 1990, Extension was a principal supporter of the Coalition for Fisheries Survival, formed in response to threats to the inshore fishery (and later the moratorium). Extension, together with the Coalition, sponsored Empty Nets, a series of community forums across the province to provide a voice for fisher people in rural outports. In 1991, the National University Continuing Education Association (NUCEA) chose Extension's Rural Women's Learning Project as the winner for creative programming in the division of programs for women. As the Evening Telegram reported, it was ironic that the award was announced on the same day the government and the university cut the Extension Service.

## Concluding

When traditional forms of organization, livelihoods, and understandings of quality of life are already under pressure, it often transpires that the forms of learning formally and informally embedded in community life and governance are inadequate. In these cases, it can be necessary to develop a truly new shared perspective on community development, and on ways of learning that can enable new (or renewed) directions. This might require a transitional governance arrangement, allowing the community to "regroup" and continue decision-making and allow the essential functioning of administration, while simultaneously creating a space to self-reflect and build assets or capacity necessary for tackling tougher challenges. Renewed participation in a process of collective self-reflection and visioning can reinvigorate the different types of learning distinguished earlier. This can lead to institutional experiments, governance reform, or a shift in local narratives. The effort could focus on restoring damaged infrastructures of learning, the quality of public discourse and education, with the eye on envisioning new perspectives at a later stage.

## References

Abelson, J., Forest, P.-G., Eyles, J., Smith, P., Martin, E., & Gauvin, F.-P. (2003). Deliberations about deliberative methods: Issues in the design and evaluation of public participation processes. *Social Science & Medicine*, *57*(2), 239–251.

Acemoğlu, D., & Robinson, J. A. (2016). Paths to inclusive political institutions. In J. Eloranta, E. Golson, A. Markevich, & N. Wolf (Eds.), *Economic history of warfare and state formation* (pp. 3–50). Springer.

Ansell, C., & Torfing, J. (2021). *Public governance as co-creation: A strategy for revitalizing the public sector and rejuvenating democracy*. Cambridge University Press.

Armitage, D. (2005). Adaptive capacity and community-based natural resource management. *Environmental Management*, *35*(6), 703–715.

Assmann, J., & Czaplicka, J. (1995). Collective memory and cultural identity. *New German Critique*, *65*, 125.

Bannister, J. (2003). *The rule of the admirals: Law, custom, and naval government in Newfoundland, 1699–1832*. Osgoode Society for Canadian Legal History.

Bellino, M. J., & Williams, J. H. (2017). *(Re)constructing memory: Education, identity, and conflict*. Sense Publishers.

Boissevain, J. (1966). Patronage in Sicily. *Man, 1*(1), 18–33.

Brenneis, D. (1988). Telling troubles: Narrative, conflict and experience. *Anthropological Linguistics*, 30, 279–291.

Brouwer, S., & Huitema, D. (2017). Policy entrepreneurs and strategies for change. *Regional Environmental Change*, 18, 1–14.

Bull, R., Petts, J., & Evans, J. (2008). Social learning from public engagement: Dreaming the impossible? *Journal of Environmental Planning and Management, 51*(5), 701–716.

Burzynski, M. (2017). *Gros Morne National Park: The most beautiful place*. Gros Morne Co-operating Association.

Carpentier, N. (2011). *Media and participation: A site of ideological-democratic struggle*. Intellect.

Cohen, A. P. (1975). *The management of myths: The politics of legitimation in a Newfoundland community*. Manchester University Press.

Crosbie, J. C. (1956). Local government in Newfoundland. *Canadian Journal of Economics and Political Science/Revue Canadienne de Economiques et Science Politique*, *22*(3), 332–346.

Crow, D. A., & Lawlor, A. (2016). Media in the policy process: Using framing and narratives to understand policy influences. *Review of Policy Research*, *33*(5), 472–491.

Dunlop, C. (2020). Policy learning and policy failure: Definitions, dimensions and intersections. In C. Dunlop (Ed.), *Policy learning and policy failure* (pp. 1–22). Bristol University Press.

Dunlop, C. A., & Radaelli, C. M. (2020). The lessons of policy learning: Types, triggers, hindrances and pathologies. In G. Capano & M. Howlett (Eds.), *A modern guide to public policy* (pp. 222–241). Edward Elgar Publishing.

Elias, N., & Scotson, J. L. (1994). *The established and the outsiders: A sociological enquiry into community problems*. Sage.

Escobar, A. (1984). Discourse and power in development: Michel Foucault and the relevance of his work to the third world. *Alternatives*, *10*(3), 377–400. https://doi.org/10.1177/030437548401000304

Fink, B. (2013). *Against understanding, volume 1: Commentary and critique in a Lacanian key*. Routledge.

Fischer, F. (2003). *Reframing public policy: Discursive politics and deliberative practices*. Oxford University Press.

Frost, F. (1958). Is artificial regeneration necessary in Newfoundland? *The Forestry Chronicle*, *34*(2), 194–200.

Gray, I., & Sinclair, P. (2005). Local leaders in a global setting: Dependency and resistance in regional New South Wales and Newfoundland. *Sociologia Ruralis*, *45*(1–2), 37–52.

Greenwood, R. (1991). *The local state and economic development in peripheral regions: A comparative study of Newfoundland and Northern Norway* [Phd, University of Warwick].

Gunder, M. (2005). Obscuring difference through shaping debate: A Lacanian view of planning for diversity. *International Planning Studies*, *10*(2), 83–103.

Held, D. (1996). *Models of democracy*. Stanford University Press.
Hillier, J., & Gunder, M. (2005). Not over your dead bodies! A Lacanian interpretation of urban planning discourse and practice. *Environment and Planning A*, *37*(6), 1049–1066.
Howlett, M. (2020). Dealing with the dark side of policy-making: Managing behavioural risk and volatility in policy designs. *Journal of Comparative Policy Analysis: Research and Practice*, *22*(6), 612–625.
Howley, K. (2009). *Understanding community media*. SAGE Publications.
Hulliung, M. (2017). *Citizen Machiavelli*. Routledge.
Jamieson, A. (2003). Gwyn, Sandra. [4.2 linear m of textual material, 1,789 photographic items, 2 audio tapes] Finding Aids of the Archives and Special Collections (COLL–316).
Jeppesen, S. (2016). *Understanding alternative media power: Mapping content & practice to theory, ideology, and political action. Democratic Communiqué*, *27*(1), Article 1.
Lankshear, C., & Knobel, M. (2011). *New literacies: Everyday practices and social learning*. McGraw-Hill Education.
Løddesøl, A. (1955). *Report on the investigation and utilization of the bogs of Newfoundland*. (p. 48). Centre for Newfoundland Studies. https://collections.mun.ca/digital/collection/cns2/id/158094/
May, P. J. (1992). Policy learning and failure. *Journal of Public Policy*, *12*(4), 331–354.
Nair, S., & Howlett, M. (2017). Policy myopia as a source of policy failure: Adaptation and policy learning under deep uncertainty. *Policy & Politics*, *45*(1), 103–118.
Newig, J., Kochskämper, E., Challies, E., & Jager, N. W. (2016). Exploring governance learning: How policymakers draw on evidence, experience and intuition in designing participatory flood risk planning. *Environmental Science & Policy*, *55*, 353–360.
Ostrom, E. (2005). *Understanding institutional diversity*. Princeton University Press.
Pahl-Wostl, C., Tàbara, D., Bouwen, R., Craps, M., Dewulf, A., Mostert, E., Ridder, D., & Taillieu, T. (2008). The importance of social learning and culture for sustainable water management. *Ecological Economics*, *64*(3), 484–495.
Sarat, A. (1998). *Everyday practices and trouble cases*. Northwestern University Press.
Scott, J. C. (1985). *Weapons of the weak: Everyday forms of peasant resistance*. Yale University Press.
Scott, T. M. (1968). The diffusion of urban governmental forms as a case of social learning. *The Journal of Politics*, *30*(4), 1091–1108.
Sider, G. M. (2003). *Between history and tomorrow: Making and breaking everyday life in rural Newfoundland*. University of Toronto Press.
Stoddart, M. C. J., & Graham, P. (2016). Nature, history, and culture as tourism attractors: The Double Translation of Insider and Outsider Media. *Nature and Culture*, *11*(1), 22–43.
The Harris Centre (Director). (2024, March 27). *Harris Centre: Media & Democracy* [Video recording]. https://www.youtube.com/watch?v=enJ6DtslbDw
Van Assche, K., Beunen, R., Verweij, S., Evans, J., & Gruezmacher, M. (2021). Policy learning and adaptation in governance; A co-evolutionary perspective. *Administration & Society, 54,* 1226–1254.
Van Assche, K., & Lo, M. C. (2011). Planning, preservation and place branding: A tale of sharing assets and narratives. *Place Branding and Public Diplomacy*, *7*(2), 116–126.
Van Assche, K., Van Biesebroeck, J., & Holm, J. (2014). Governing the ice. Ice fishing villages on Lake Mille Lacs and the creation of environmental governance institutions. *Journal of Environmental Planning and Management*, *57*(8), 1122–1144.

Wadel, C. (1968). *Marginal adaptations and modernization in Newfoundland: A study of strategies and implications in the resettlement and redevelopment of outport fishing communities*. Institute of Social and Economic Research, Memorial University of Newfoundland. https://cir.nii.ac.jp/crid/1130282271615318016

Wash, I. (2020). Interpreting public policy dilemmas: Discourse analytical insights. *Humanities and Social Sciences Communications*, *7*(1), Article 1.

Webb, J. A. (2008). *The voice of newfoundland: A social history of the broadcasting corporation of Newfoundland, 1939–1949*. University of Toronto Press.

Whelan, M. (2002). *The newspaper press in nineteenth-century Newfoundland: Politics, religion, and personal journalism* [Doctoral, Memorial University of Newfoundland]. https://research.library.mun.ca/1490/

Yanow, D. (1992). Silences in public policy discourse: Organizational and policy myths. *Journal of Public Administration Research and Theory*, *2*(4), 399–423.

# 4 Governance, experiment, and reform for rediscovering strengths

**Abstract**

*Governance is a process of collective decision-making, enabling collective action and coordination, unique to each community. At the same time, it is an outcome: the structures that form where that decision-making takes place, the resulting procedures, tools, and arenas, and patterns of inclusion and exclusion. Governance differs from government, but generally, a variety of governmental actors participate in governance and can rely on formal or informal tools for coordination.*

Whereas stories and learning are in the realm of thinking as and about the collective, governance is about organizing. The way a community understands itself will determine the way it organizes itself, both in the present and anticipating the future (Czarniawska & Gagliardi, 2003; Elwood, 2006). The policy tools and forms of organization a community uses and chooses to improve also hinge on the stories it tells itself, and the values and feelings that come with those stories. The forging of new stories can take place within the governance system or in the wider community, via public discourse and learning. Existing and new stories can be connected to policy tools, which might turn these stories into a strategy for the future. The creation of new assets, reinterpretation of old assets, and choices for new collective futures take place through governance (Van Assche et al., 2023).

Our perspective on governance is evolutionary, meaning that to comprehend the way things are organized in a community implies understanding the way governance and community co-evolved. Ways of organizing are shaped by a history of articulating and using certain policies, forms of organization, and procedures, which have given some people and stories a stronger voice than others and defined problems and opportunities in particular ways (Van Assche, Verschraegen, et al., 2021). Rethinking what assets could be and how they might contribute to future strategies takes place in a context that is

DOI: 10.4324/9781003620297-4
This chapter has been made available under a CC-BY-NC-ND 4.0 license.

shaped by the past, including old ideas and old forms of organizing, but it is not uncommon that serious reflection on new assets and futures will find that governance *reform* is needed.

Community reflection and assessments may determine that it is not possible to form a collective vision of a better collective future, or to organize for a desired future because existing decision-making mechanisms do not work properly or allow for such reflection or imagination (Howlett & Ramesh, 2014). Each community defines good governance in its own way, but many face an impending system where the governance infrastructures to move in a direction preferred by communities are missing. In these cases, a preference for a particular future cannot be articulated because existing governance structures do not allow for it (Cohen, 1975b; Lodge & Wegrich, 2014).

If governance reform seems warranted but does not appear possible, the experiment might be an alternative. Experiments can be designed in various ways, depending on their scope, context, and the issues responded to (Nair & Howlett, 2016). Much like institutional capacity, the utility of an experiment depends on the context within which it takes place, including the community employing it and the governance system it is attempting to improve. The experiment might address an institutional weakness or missing capacity, aim to solve an old problem in a new way, or involve total governance reform in a specific area.

Newfoundland and Labrador (NL) have a strong tradition of governance experiments, some of which are featured as vignettes in this book. Since smaller communities that depend on natural resources, relatively isolated, and/or equipped with relatively under-developed governance systems can be found on every continent, our book is partly motivated by the desire to share these experimental histories so other communities can benefit (Butters et al., 2018; Hall et al., 2017).

## Governance: actors and institutions

Each governance system is the result of history, with certain forms of participation and representation practiced and accepted over time. Rules can be written or unwritten, gatherings can be regular or motivated by external events, patterns of participation can shift, and different circles may focus on different topics, associated with different roles or families in the community (Friedmann, 1992). Governance systems can develop extensive administrations (or not), and they can look far into the future or be more focused on the present and past. Leadership can be concentrated or more widely distributed, and it can be elected, emerge through other forms of competition, or can be associated with cultural or economic status, age, gender, profession, religion, lineage, or ethnicity.

Governance engages a variety of actors, producing and relying on policies, plans, laws, and informal institutions and structures at different scales (Beunen et al., 2015; Marquardt, 2017). Actors are those individuals, groups, and associations that directly affect decision-making, meaning that they could be around the table when it happens, or they could be in different settings where ideas are formulated, considering possible futures, threats, and opportunities (Beritelli & Laesser, 2011; Fischer & Newig, 2016). We label as actors those who have agency and intention, who are recognized as a unit capable of expressing a common point of view. Actors and their roles are never entirely static. They can disappear, meaning their unity, point of view, and agency can vanish. Their role can shift, meaning that they can become more or less powerful, shift their scope of interest, or change their focus of intervention (Craps et al., 2019).

Actors come together in certain *arenas* to decide on issues they deem relevant and pressing for the community. They use and develop rules of coordination, or institutions (Ostrom, 2005). Action is coordinated by *institutions* and making institutions is structured by other institutions – rules and traditions that define places, times, and procedures for the production and application of new rules. Governance shapes the collective, and not just anyone can decide over matters that will affect the whole, certainly not in random ways. In the process of governance, stronger bonds emerge between certain actors, shaping power relations, and some stories about the community, livelihoods, and quality of life become more important than others (Garud & Karnoe, 2001).

In complex societies, individuals exert influence often via an organization; businesses, cooperatives (see Fred Campbell's vignette below) administrative departments, non-profits, or cultural or religious organizations, and others are thus all potential governance actors (Besio & Pronzini, 2011). Therefore, understanding governance requires understanding the internal governance structures of such relevant organizations. Groups can organize themselves and develop internal governance structures and procedures. A religious group, parish, youth council, or collection of people sharing a specific concern or aligned along natural landscape boundaries (e.g., a watershed or coastline) are all examples of groups with their own governance structures, increasing their capacity to participate in community governance (Allen & Cochrane, 2010). Participation and representation can come about, as well as tools of coordination and patterns of inclusion and exclusion. Groups can govern themselves largely outside the sphere of community governance or exert a more marked influence. They can slowly take over governance when collective decisions are *de facto* taken by a group which represents only selected interests (Mahoney, 2004).

## For the people: co-ops, adult education, and the Antigonish movement

Fred Campbell

Credit unions have a long history in NL, going back to the Commission of Newfoundland Government (1934/1949), which during the global depression replaced the rule of an independent democracy with six commissioners chosen by the British government. Cooperatives go back even further, to the so-called Antigonish movement, associated with St. Francis Xavier University (St. FX) in Antigonish, Nova Scotia. The movement, championed by Father Moses Coady, promoted a broadening of education beyond the elites. By 1930, St. FX had established a Department of Extension, or Department of Social Action with the "social vision" of a cooperative society and programs for fishermen, miners, lumbermen, women, and other "ordinary people" in the time of the Depression.

Cooperatives were seen to provide a natural schooling in local organizations and connections were formed in Atlantic Canada between credit unions, adult education, and cooperative organizations, enhancing their role in local self-organization and community development. Moses Coady in his 1939 book, *Masters of Their Own Destiny*, goes on to explain that Newfoundlanders organized

> for study and economic action according to the Antigonish technique. Their success in the face of many obstacles was encouraging. In 1938 there were six hundred study clubs, twenty-six credit unions, twenty-five buying clubs, and ten other cooperatives, A program of producer and consumer cooperation is gradually being established in Newfoundland.
>
> (p. 67)

The influence of the cooperative movement in Newfoundland can still be felt. Five decades ago, according to local author Roy Dwyer, Fogo Island was saved from resettlement by the creation of a fishery cooperative, the Fogo Island Co-operative Society Limited, which still operates today. In 2024, a new Credit Union opened in Fogo, replacing the Bank of Nova Scotia, which continues to close its rural branches in rural Atlantic Canada.

In 1949, when Newfoundland became a province of Canada, first premier Joey Smallwood upgraded Memorial University of Newfoundland (MUN) from a college and hoped that Memorial would immediately establish a Department of Extension, modeled on the lines of the Department at St. FX. The connection of St. FX continues as Newfoundlanders and Labradorians with an interest in adult education and community development continue to enter master's programs at the home of the "Antigonish Movement."

---

Formalizing governance means choosing predictable times and places to meet, repeatable procedures of discussion, deliberation, and decision-making, with standardized forms of participation and representation. What formalization means will be unique for each community and so will be the balance between formality and informality (Guha-Khasnobis et al., 2007). Small communities can be dominated more easily by a family, profession, or grand leader, yet chances are that this only works with the tacit support of other individuals, organizations, or higher-level actors. Informal networks and institutions can be layered and opaque, yet their existence represents a form of governance. It also entails that over-ambitious individuals have a hard time bending the community according to their will (Casson et al., 2010). The formal structures which developed in nation states, and which we recognize as the apparatus of government, provide a set of visible actors and institutions, ideally stabilized and representative, yet their importance in the whole configuration can vary dramatically. Governance can function without strong governmental actors, but the presence of predictably designated leaders and stable administration can increase the capacity of a community to organize itself and its future.

Investigating who has a role in governance and who doesn't certainly raises the question who should have a role or who should not. This directly leads to issues of participation and representation and the importance of observation in identifying patterns of inclusion and exclusion (Karambelkar & Gerlak, 2020). Opening new avenues of participation can open thinking in governance about assets and futures, but as local leaders and citizens, as external advisors or simply as observers, we must be careful not to assume that correct or fair patterns of participation are easy to recognize (Cooke & Kothari, 2001; Pløger, 2021). If patterns of exclusion are not taken into consideration in collectively defining a community's future, this might render the exercise ineffective.

## Strategy and reform

Community development requires governance both for rethinking and for organizing desirable futures. Within governance, assets and their possible futures can be articulated, compared, and assessed, aided by the learning processes discussed earlier. Both creating new stories and connecting these stories to old and new policies in developing strategy must find a place in governance (Van Assche et al., 2023).

Negative experiences or no experience with rethinking assets and building strategy can often inspire apathy. If a highly participatory process to rethink assets required people investing a lot of effort, and if expectations were high but afterward nothing happened, or sudden bureaucratic hurdles or local infighting took place, a new process to rethink the possibilities for community development may not find much traction. This means that less participation can be expected compared to earlier processes and that ideas will be less representative, that less persuasive stories will connect less, and that the vision being pursued is less likely to become a reality.

Often a community trying to reconsider its assets comes to the realization that governance requires some repair or upgrading. Considering that governance is more than government, and governance always exists in some multi-level arrangement, governance reform must envision more options than administrative or political reform, and it must consider different possible levels of intervention (Moss & Newig, 2010; Papadopoulos, 2007). We reiterate that trying to copy ideal models of local or regional governance, or “good” government or governance, is highly problematic, since somethings work for certain people in certain environments but not necessarily for others. Reform will be shaped by what is possible, what is seen as desirable locally, by the stories that frame our thinking about the good community and good governance. Advisors, politicians, or academics may present simple solutions as ideal, often because of assumptions and stories that are not made explicit. For example, when radical participation is proposed as the solution to all issues, the motivation of those proposing it might be rooted in experiences of extreme exclusion, in a felt urgency to give voice to the oppressed.

Governance reform can alter relations between the council and administration, politics and civil society, or economic players and the local community (Long, 1949). Reform can follow the ideology of new public management by focusing on simplification and precise definitions of roles, responsibilities, and products. It can also go in the other direction, enabling actors to see the bigger picture and allowing for an institutional toolbox, which is big enough to solve new problems, shift the general direction, and interrogate the big picture. Once again, what ought to be reformed hinges on the actual problem, as well as the desire of the community and its own aspirations and understandings of good governance (Gunder & Hillier, 2009; Hood & Peters, 2004; Van Assche, Verschraegen, et al., 2021). For us most relevant is the fact that governance reform might enable a community to rethink itself, and, from that new starting point, chart a new course. Increasing institutional capacity, meaning the capacity to decide on more issues and to organize more aspeccts of life, will often appear as central, yet upon reflection, a community might decide to focus on very specific and seemingly minor reforms which nonethelss unlock a rethinking of assets.

Governance reform might feel like an ambitious challenge for many small communities, one that many may not feel confident to take. Transitional governance arrangements might become an opportunity for the community to deploy transitional forms of leadership, to break away from unhealthy dependencies, or to manage specific risks (Van Assche, Duineveld, et al., 2021). Transitional leadership and governance are meant to be temporary, acting as a controlled and temporary disrupter, eventually fading away while allowing a new arrangement of actors to consolidate. They can stabilize the community and offer opportunities to rethink and set the ground for strategizing (Van Assche et al., 2023). Rethinking assets could transpire within the context of such an arrangement; transitional governance could be expressly created as a space for self-interrogation. Another possibility is to create transitional governance arrangements in preparation for governance reform, a space to reflect on how to embark on reforms, or an opportunity to craft a reform strategy. Such possibility, however, is more promising when a basic agreement on key strengths is already in place.

The story of NL demonstrates how difficult it is to conduct radical governance reform, for example, the establishment of regional government. Both insiders and outsiders have recognized for several generations that the weak and fragmentary municipal structures and the absence of regional governance creates difficulties, both for basic service provision and for more abstract community development goals (Bannister, 2003; Crosbie, 1956; Powell, 1949). Either stronger local government or a new form of regional government has been recognized as desirable for increasing community capacity to rethink and reorganize, yet reform in either of those directions has proven daunting.

The outports before confederation, prior to learning about municipal forms of organization elsewhere in the country, were of course not unorganized. Although largely ignored, and undermined, by colonial powers, Indigenous peoples and communities had their own forms of governance prior to and beyond European settlement. Hudson (2020), for example, describes decision-making guided by "Inuit values respecting reverence for culture, tradition, family, health, safety, respect for all relations (human and natural environment), and responsibility to community" (pp. 51–52). Velez introduced Mi'kmaq governance concepts in Chapter 1, such as *Netukulimk* and *nmitkinen aqq msit ko'kmanaq* (our territory and all our relations) and the *kloqowej* model for Mi'kmaw law and justice, that continue to guide understandings and goals related to community well-being and sustainability based on peaceful, harmonious relationships between Mi'kmaq and the land as well as values of sharing, respect, reciprocity, and responsibility for the future (Figure 4.1).

*Figure 4.1* Amy Hudson enjoys a sunset at St. Lewis, Labrador. Credit: Andrea Procter.

In outport settlements, regardless of their diverse histories, people came to identify with the place, its activities, and a set of values and collectively binding decisions in their communities and regions. What gradually began to emerge in most areas was a situation where relatively small municipalities began to bear the weight, in terms of service provision, of surrounding unincorporated areas. Once an idea of minimal services, minimal consideration of alternative futures, and minimal taxes was established, such forms of organization (for example those associated with local service districts and unincorporated areas) became associated with local identity (Overton, 1995; Webb, 1998). Ideas of either amalgamation with neighbors or full incorporation became seen as unreasonably costly (it was expected taxes would rise or a new municipal entity would focus on things that were unnecessary) and also perceived as an attack on local identity and autonomy. If local government does not offer arenas to consider alternative futures, assessing the need for a different form of local government where such functions might work is challenging. Inclusive of all communities in a region, Regional Economic Development Boards (REDBs) offered such an arena during their time of existence (see the vignette by Rob Greenwood below).

Resistance to change, as seen in many places beyond NL as well, often translates into voters identifying and electing politicians who maintain the *status quo*. Initiatives to amalgamate municipalities, force incorporation, or enforce the creation of regional governance thus encounter a formidable obstacle. Resistance to amalgamate or incorporate is also tied to memories of resettlement, with people moving to places where governments could more easily provide services in a larger municipality perceived and experienced by many as losing local identity (Cohen, 1975a; Matthews, 1979; Overton, 2000).

Regional government or governance could be a way to reinforce the lobbying power of local governments and communities and strengthen services and checks and balances, while at the same respecting local identities. Regional arenas can allow communities to compare notes, experiences, or ideas, encouraging comparative learning, sharing of expertise, and new experiments at local levels. They can also allow for collaboration or coordination in policy arenas where regional support is found. Regional arenas can also enhance dialectical learning, which in turn can entail rethinking assets or building different futures from existing assets (Butters et al., 2018; Hall et al., 2017). Connecting ideas to resources and policy tools, including local but also external tools, can often happen more easily within larger arenas. We turn next to governance experiments, including NL's history of experimentation with regional governance – but not government (see also Vodden et al., 2013).

## Governance experiments

As described by Campbell above, in the 1920s, Jesuits from Nova Scotia, associated with the St. FX University in Antigonish, promoted cooperative forms of organization in Newfoundland. The movement promoted collaboration in ship building, fishing, and retail to give communities more income but also more agency, and a slightly more democratic distance from dominant merchants. Communities gained experience in self-organization and self-transformation, even in the absence of local governments (Coady, 1939; Plunkett, 1955).

Between the 1950s and 1970s, Memorial University's Extension Service built on this tradition, combining it with distance learning. Project workers attempted to empower local communities and promote self-organization through the formation of cooperatives but also local arenas for discussing shared problems and interests and, where possible, to envision alternative futures (MacLeod, 1990). Communities were encouraged to take hold of their own destiny, and Extension Workers tried to connect this with knowledge, things that might be useful to learn, resources, and ways of financing new-found ambitions. Newhook describes below one such effort, The Fogo Process. Sometimes, Extension activities led to formal reform, such as the creation of municipalities or the amalgamation of formerly competing communities (Newhook, 2009).

---

**The Fogo Island Film Project and its secret weapon**
**A self-reflection process that changed the future and development path of several communities**

Sue Newhook

In the summer of 1967, a famous documentary filmmaker from the mainland showed up on Fogo Island, and he and his crew made a bunch of films that saved the island from resettlement. Or so the story sometimes goes....

It's true the National Film Board's Fogo Island project used film as a tool of community development to help jumpstart a co-op that kept islanders from moving away, and that the project set a template for what would become internationally known as the Fogo Process. To define the process simply: people in isolated communities discussed community issues on camera, and the resulting films sparked more discussion in other communities, building consensus and cooperation. True, too, that at the project's end, Fogo Islanders from a dozen settlements started a fisheries cooperative that still helps to support the island today. What this simple description misses is that the National Film Board (NFB) entered a process that had already started. It gave a final push to the work of a determined group of citizens and a local development worker whose job was to stay in the background. Forty years later, the development worker's own files helped bring his pivotal role to light.

Fred Earle worked for the Memorial University Extension Service, which led to many social and cultural changes in NL throughout the 1960s and 1970s. He was to support and create conditions for grassroots development, rather than setting agendas himself. He brought a lifetime of connections to his first day of work in 1964 at the age of 50. Born and raised in nearby Change Islands, he knew everyone and could talk to anyone; he was a relentless networker. With gentle but persistent nudging from Earle, islanders had created the Fogo Island Improvement Committee. It had made some progress lobbying for government attention but needed more support from their neighbors and the government in St. John's. Earle's boss wanted the NFB to shine the light of its new national project on Fogo Island. Don Snowden had a background in northern co-op development and a range of connections in Ottawa. When he heard the NFB wanted to make some activist development films for a Canada-wide program called Challenge for Change, he wanted in. He arranged to take NFB documentarian Colin Low on a tour of several Newfoundland outports, saving Fogo Island for last.

As Snowden hoped, Low found the story and the partners he was looking for: "a community in trouble… whose people had already begun to organize for change, so we could affect the organization process by improving communication." Low knew that "as outsiders, (they) could never go into a community without the help of such a person" as Fred Earle (Low, 1968). Sure enough, many people wouldn't talk to the mainlanders at first, but with Earle's help, by early 1968, a mix of 27 short films had been screened, discussed, and argued about in communities around the island. In early December 1967, islanders met to create a fisheries cooperative that modernized their fishing fleet and took more control of their catch from outside interests.

---

Rural development associations go back to 1968, beginning as a reaction to governmental policies of industrialization and resettlement. In the 1980s, regional development boards, a further developed version of the initial associations, became active in bringing together private and public partners to encourage economic development (Vodden et al., 2014), as described by Rob Greenwood below. It was only in 1998 that municipalities were able to become more independent, setting their own economic development goals with the support of the province.

The rural development associations (and later REDBs) left traces in the governance landscape, as they enabled communities to organize themselves on issues previously out of reach, and the learning process in some places consolidated local governments, led to follow-up initiatives, or more concerted efforts to make provincial policy sensitive to local concerns (Hall et al., 2017; Vodden et al., 2014) (Figure 4.2).

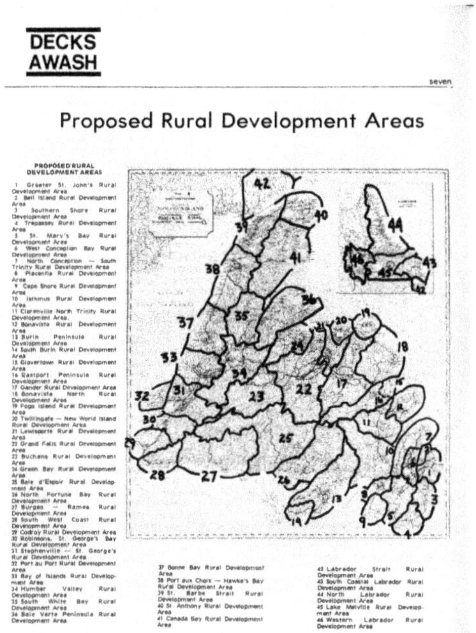


*Figure 4.2* Illustration from Memorial University's Extension service newsletter *Decks Awash* proposing different rural development areas in Newfoundland. Memorial University Extension Service (1971, 10) Decks Awash, vol. 1, no. 08 (October 1971) [magazine/text] Memorial University of Newfoundland. Libraries. Centre for Newfoundland Studies (Decks Awash, vol. 01, no. 08 (1971)).

In St. Anthony, on the tip of the Northern Peninsula of Newfoundland, St. Anthony Basin Resources Inc. (SABRI) was established in 1997 to manage a new shrimp quota. SABRI worked with the town, the port authority, and others to use profits from shrimp for economic development and diversification. They used these revenues to build cold storage and improved harbor facilities, not only for seafood export but also in the hope of attracting larger container carriers and cruise ships. Beyond this, SABRI has invested in social housing, building tourism infrastructure, and transportation infrastructure to make existing assets more accessible (e.g., trails, bus service) (Johnson, 2002; Lowery et al., 2023).

Through comparative research with a development partnership in rural Ireland, Gibson (2014) provides yet another example of governance learning and experimentation, examining the experience of the relatively short-lived Northern Peninsula Regional Collaboration Initiative, created by the province in 2009. Gibson's research illustrates the financial and human investments required to support governance experimentation as well as the need for new mechanisms within government to work with communities and regions as partners, for building trust and relationships, and for evaluating collaborative governance processes in new ways.

Meanwhile, the idea of regionalization has never been abandoned in the province. The dissolution of REDBs (see below) left many communities with a desire to find new forms of collaboration and valuable learning experiences. Municipalities NL played a double role of encouraging social innovation, administrative reform, and governance experiments on the one hand and providing an arena to discuss local, regional, and provincial issues and the functioning of the multi-level governance system on the other. Despite another apparently failed attempt in 2017–2022 to initiate regional government in the province, discussions continue about what future forms of regional governance might look like.

---

**Regional Economic Development Boards in Newfoundland and Labrador governance experiments and reform: getting strategy and structure right, but failing to understand the power of political culture?**

Robert Greenwood

NL initiated a process to establish REDBs in 20 Economic Zones in January 1995, with the publication of the report of the Federal-Provincial Task Force on Community Economic Development (CED). In May 2012, the federal government announced it was discontinuing funding to the REDBs and the provincial government quickly followed. This resulted in an era with no regionally based community or regional development capacity in most of the province, which continues to the present.

Canada, according to the OECD, has the weakest local government of all its members. Provinces are very strong relative to the federal government and municipalities are, under the constitution, "creatures of the provinces." Municipalities in NL reflect this reality and are also the product of the province's history as a fishing outpost, where formal governance powers were slow to develop, and municipal government did not spread even after NL joined Canada in 1949. In rural outports, especially, power and authority had been wielded by the church, merchants, and far away colonial masters, and after confederation, by politicians and bureaucrats in St. John's and Ottawa.

Community and regional economic development are not solely reliant on local government, of course. NL had cooperatives, the Fisheries Protective Union, and the work in northern NL of Sir Wilfred Grenfell. When the first provincial government, led by Joey Smallwood, initiated the resettlement program, a network of Regional Development

Associations (RDAs) emerged, initially to oppose resettlement. Memorial University's Extension Service (MUN Extension) played a key role in supporting this process, providing leadership development and the fundamentals of organizational development.

Over time, the federal and provincial governments provided support to RDAs to hire coordinators and implement development projects in sectors like fisheries, agriculture, and tourism. Provincial government field staff worked alongside MUN Extension fieldworkers to support development projects and capacity building. As the unemployment insurance system was extended to fishermen and other seasonal workers, RDAs and rural municipalities became key vehicles to administer short-term "make work" projects to enable seasonal workers to qualify for unemployment insurance. This provided essential income support, funded through the federal government rather than resorting to "welfare," funded by the province but RDAs became seen by many, including the federal and provincial governments that funded them, as "stamp factories," generating projects with little long-term contribution to economic development.

By the late 1980s, there were close to 60 RDAs in NL, complemented by Community Futures Committees established by the federal government nationally to develop strategic plans for their regions, support capacity building, and promote economic development and diversification. By 1989, there were 13 (later 17). The program also supported the creation of 13 (eventually 16) Business Development Centers to provide small business loans. This institutional diversity was further enriched – or complicated – by the creation of the federal Atlantic Canada Opportunities Agency in 1987, with its own field staff, the Women's Enterprise Network, federal-provincial telecenters, and various development corporations created as pilots inspired by the 1986 report of the Royal Commission on Employment and Unemployment (RCEU) and as adjustment efforts in the face of the burgeoning fishery crisis.

In 1989, the provincial government created the Economic Recovery Commission (ERC), chaired by Dr. Doug House. The umbrella organization for municipalities in the province asked Dr. House to convene a meeting with the umbrella for RDAs to explore possible collaboration. House offered me a position to facilitate the process. The political pressure to address the plethora of seemingly competing CED organizations, in a time of resource crisis and fiscal restraint, resulted in the provincial premier, Clyde Wells, drawing 17 economic zones on the map and asking House to work with his federal equivalent at Atlantic Canada Opportunities Agency (ACOA) to create a new coordinated approach. The federal-provincial Task Force on CED in NL was the result. I was assigned to lead research and analysis, serve as principal author, and then lead implementation through a federal-provincial-community Economic Zone Unit.

I and others have written about the establishment, operation, and evaluation of the REDBs and their eventual demise. More is to be written. I believe the Task Force and the hundreds of volunteers and staff who worked throughout the province on the provisional boards and the REDBs, and those who supported their efforts forged a system that successfully established strategic development priorities for all regions of the province, advanced leadership development and community capacity building, and resulted in many regional and sectoral economic development successes.

On reflection, as someone whose career, and life, has been inextricably linked to the study and public policy development around community and regional economic development in NL, I think we got a lot right about the strategy and structures required for success. Where I think we failed was in understanding the political culture of NL, especially in rural regions. Peter Drucker's famous line that "culture eats strategy for breakfast," is correct. Citizens in Norway have 1,000 years of experience in local democratic governance. But because of their history, Newfoundlanders and Labradorians do not have a strong faith in the efficacy of formal democratic institutions with the power and resources to make decisions that impact your life but are seen to reside outside your influence. Formal governance at the local level – such as planning rules and regulations impacting access to the land, inland waters, or the sea for food, home construction, and other uses – is often seen as counter to the culture. Simultaneously, efforts to strengthen regional self-governance have likely been seen as a threat by provincial politicians, whose power depended on being the primary route to government programs and funding, contributing to centuries of top-down paternalism and patronage, as well as bottom-up defiance. Indigenous communities in NL, of course, have very different traditions and histories, and they have seen significant progress in terms of building capacity for community and regional economic development.

## Concluding

Governance is an essential infrastructure for reflecting on assets and crafting community strategy based on such reflection. In governance, narratives about the community can emerge, be introduced, or reconfigure. They can become accepted for guiding future development and translated into strategies. Rethinking assets is of little use if there is no possibility to come to a shared interpretation and no way to act collectively on such interpretation.

Governance must be understood and mobilized to rethink assets and futures. What we discovered in this and previous chapters is that this harnessing is not a simple matter and that the infrastructure might need some repair before one can fruitfully develop a new perspective on community futures. Repair might be an outright reform, meaning a long-term commitment to change and rethinking of a wide range of topics and places, structures, and procedures. It can also take the form of an experiment. In NL, the need for governance change has often been felt yet reform has been difficult, leaving experiment as a natural path toward change.

The NL experience also illustrates some key points regarding the forms of governance that enable rethinking long-term futures and assets.

*First*, it illustrates that "good governance," where the aim is reframing, adapting, and possibly reinventing, requires multi-level structures where mutual correction and support can play out and where checks and balances can develop, so no perspective can easily override another and various perspectives and realms of responsibility can be considered. *Second*, we see that governmental reform without a fuller understanding of governance is doomed, but conversely, governance systems without governmental support make those systems vulnerable and weaken their institutional capacity. *Third*, strong local identities can easily survive new forms of collaboration, as well as development of more formal local governments. *Fourth*, strong identities, attachment to place, and lifestyles can adapt to realize that both thinking and organizing might need to change to sustain what is most valuable to communities. *Finally*, we have seen that experiments might be discontinued but still leave productive traces both in the places where the experiment took place and in other communities that took note. Thus, such experiments can inspire learning and lay the groundwork for future reform.

Governance cannot be entirely designed, but its evolution is also not a product of chance. Governance paths create their own structure, variation, and flexibility. What and how paths can be changed to facilitate the articulation of alternative assets and futures will be different in each case (Van Assche et al., 2014). No recipe for governance reform or strategizing within governance can work in predictable ways. What is key is a willingness in the community to examine both the strengths and the limitations of the current self-understanding and governance, leaving space to consider governance learning and change where it is needed.

## References

Allen, J., & Cochrane, A. (2010). Assemblages of state power: Topological shifts in the organization of government and politics. *Antipode*, *42*(5), 1071–1089.

Bannister, J. (2003). *The rule of the admirals: Law, custom, and naval government in Newfoundland, 1699–1832*. Osgoode Society for Canadian Legal History.

Beritelli, P., & Laesser, C. (2011). Power dimensions and influence reputation in tourist destinations: Empirical evidence from a network of actors and stakeholders. *Tourism Management*, *32*(6), 1299–1309.

Besio, C., & Pronzini, A. (2011). Inside organizations and out. Methodological tenets for empirical research inspired by systems theory. *Historical Social Research/Historische Sozialforschung, 36*(1), 18–41.

Beunen, R., Van Assche, K., & Duineveld, M. (2015). *Evolutionary governance theory: Theory and applications*. Springer.

Butters, L., Okusipe, O. M., Eledi, S. B., & Vodden, K. (2018). Engaging the past to create a new future: A comparative study of heritage-driven community development initiatives in the Great Northern Peninsula. *Journal of Rural and Community Development, 12*(2–3), 186–209.

Casson, M. C., Della Giusta, M., & Kambhampati, U. S. (2010). Formal and informal institutions and development. *World Development, 38*(2), 137–141.

Coady, M. M. (1939). *Masters of their own destiny; The story of The Antigonish movement of adult education through economic cooperation.* Harper & Brothers Publishers.

Cohen, A. P. (1975a). The definition of public identity: Managing marginality in outport Newfoundland following Confederation. *The Sociological Review, 23*(1), 93–119.

Cohen, A. P. (1975b). *The management of myths: The politics of legitimation in a Newfoundland community*. Manchester University Press.

Cooke, B., & Kothari, U. (2001). *Participation: The new tyranny?* Zed Books.

Craps, M., Vermeesch, I., Dewulf, A., Sips, K., Termeer, K., & Bouwen, R. (2019). A relational approach to leadership for multi-actor governance. *Administrative Sciences, 9*(1), 12. https://doi.org/10.3390/admsci9010012.

Crosbie, J. C. (1956). Local government in Newfoundland. *Canadian Journal of Economics and Political Science/Revue Canadienne de Economiques et Science Politique, 22*(3), 332–346.

Czarniawska, B., & Gagliardi, P. (2003). *Narratives we organize by* (Vol. 11). John Benjamins Publishing.

Elwood, S. (2006). Beyond cooptation or resistance: Urban spatial politics, community organizations, and GIS-based spatial narratives. *Annals of the Association of American Geographers, 96*(2), 323–341.

Fischer, L.-B., & Newig, J. (2016). Importance of actors and agency in sustainability transitions: A systematic exploration of the literature. *Sustainability, 8*(5), 476.

Friedmann, J. (1992). The past in the future: History and the politics of identity. *American Anthropologist, 94*(4), 837–859.

Garud, R., & Karnoe, P. (2001). *Path dependence and creation*. Psychology Press.

Gibson, R. (2014). *Collaborative governance in rural regions: An examination of Ireland and Newfoundland and Labrador* [Doctoral, Memorial University of Newfoundland]. https://research.library.mun.ca/6468/

Guha-Khasnobis, B., Kanbur, R., & Ostrom, E. (2007). *Linking the formal and informal economy. Concepts and policies*. Oxford University Press.

Gunder, M., & Hillier, J. (2009). *Planning in ten words or less. A Lacanian entanglement with spatial planning*. Ashgate.

Hall, H. M., Vodden, K., & Greenwood, R. (2017). From dysfunctional to destitute: The governance of regional economic development in Newfoundland and Labrador. *International Planning Studies, 22*(2), 49–67.

Hood, C., & Peters, G. (2004). The middle aging of new public management: Into the age of paradox? *Journal of Public Administration Research and Theory, 14*(3), 267–282.

Howlett, M., & Ramesh, M. (2014). The two orders of governance failure: Design mismatches and policy capacity issues in modern governance. *Policy and Society, 33*(4), 317–327.

Hudson, A. (2020). *Reclaiming Inuit governance: Planning for a sustainable future in Nunatakavut* [Doctoral, Memorial University of Newfoundland]. https://research.library.mun.ca/14950/

Johnson, M. L. (2002). *Decline and adjustment at the periphery, West Coast of Newfoundland.* Institut national de la recherche scientifique.

Karambelkar, S., & Gerlak, A. K. (2020). Collaborative governance and stakeholder participation in the Colorado River basin: An examination of patterns of inclusion and exclusion. *Natural Resources Journal, 60*, 1.

Lodge, M., & Wegrich, K. (2014). *The problem-solving capacity of the modern state: Governance challenges and administrative capacities*. Oxford University Press.

Long, N. E. (1949). Power and administration. *Public Administration Review, 9*(4), 257–264.

Low, C. (Director). (1968). *An introduction to Fogo Island.* National Film Board of Canada.

Lowery, B., Langer, S., Stoddart, M., Elliott, S., & Vodden, K. (2023). Linking divergent perspectives and stakeholders: A case study of St. Anthony Basin Resources Inc. In *Revitalizing PLACE,* (pp. 78-108). Memorial University Press.

MacLeod, M. (1990). *Bridge built halfway: A history of Memorial University College, 1925–1950.* McGill-Queen's Press.

Mahoney, C. (2004). The power of institutions: State and interest group activity in the European Union. *European Union Politics, 5*(4), 441–466.

Marquardt, J. (2017). Conceptualizing power in multi-level climate governance. *Journal of Cleaner Production, 154*, 167–175.

Matthews, R. (1979). The smallwood legacy: The development of underdevelopment in Newfoundland 1949–1972. *Journal of Canadian Studies, 13*(4), 89–108.

Moss, T., & Newig, J. (2010). Multilevel water governance and problems of scale: Setting the stage for a broader debate. *Environmental Management, 46*(1), 1–6.

Nair, S., & Howlett, M. (2016). Meaning and power in the design and development of policy experiments. *Futures, 76*, 67–74.

Newhook, S. (2009). The godfathers of Fogo: Donald Snowden, Fred Earle and the roots of the Fogo Island Films, 1964–1967. *Newfoundland Studies, 24*(2), 171–197.

Ostrom, E. (2005). *Understanding institutional diversity*. Princeton University Press.

Overton, J. (1995). Moral education of the poor: Adult education and land settlement schemes in Newfoundland in the 1930s. *Newfoundland and Labrador Studies, 11*(2), 250–282.

Overton, J. (2000). Academic populists, the informal economy and those benevolent merchants: Politics and income security reform in Newfoundland. *The Journal of Peasant Studies, 28*(1), 1–54.

Papadopoulos, Y. (2007). Problems of democratic accountability in network and multilevel governance. *European Law Journal, 13*(4), 469–486.

Pløger, J. (2021). Politics, planning, and ruling: The art of taming public participation. *International Planning Studies, 26*(4), 426–440. https://doi.org/10.1080/13563475.2021.1883422.

Plunkett, T. (1955). *Local government in greater Corner Brook, Newfoundland—ProQuest* [McGill University]. https://www.proquest.com/docview/2665137864?pq-origsite=gscholar&fromopenview=true&sourcetype=Dissertations%20&%20Theses

Powell, C. W. (1949). Problems arising from a lack of organized municipalities in Newfoundland. In P. T. Clark (Ed.), *1949 Proceedings of the First Annual Conference of*

*The Institute of Public Administration of Canada: Held at the Chateau Frontenac, Quebec, October 3rd and 4th* (144764.cpd). Institute of Public Administration of Canada; Memorial University of Newfoundland. Centre for Newfoundland Studies. http://collections.mun.ca/cdm/ref/collection/cns2/id/143619oai:collections.mun.ca:cns2/143619

Van Assche, K., Beunen, R., & Duineveld, M. (2014). Formal/informal dialectics and the self-transformation of spatial planning systems: An exploration. *Administration & Society*, *46*(6), 654–683.

Van Assche, K., Duineveld, M., Gruezmacher, M., & Beunen, R. (2021). Steering as path creation The art of managing dependencies and reality effects. *Politics and Governance, 9*(2), 369–380. https://doi.org/10.17645/pag.v9i2.4027.

Van Assche, K., Gruezmacher, M., Marais, L., & Perez-Sindin, X. (2023). *Resource communities: Past legacies and future pathways*. Taylor & Francis.

Van Assche, K., Verschraegen, G., & Gruezmacher, M. (2021). Strategy for collectives and common goods. Coordinating strategy, long term perspectives and policy domains in governance. *Futures*, *128*. https://doi.org/10.1016/j.futures.2021.102716.

Vodden, K., Gibson, R., & Daniels, J. (2014). Newfoundland and Labrador provincial regional development policy. *Canadian Regional Development*, *18*, 31.

Vodden, K., & Hall, H. with D. Freshwater and Research Team (2013). Understanding regional governance in Newfoundland and Labrador: A survey of regional development organizations. Prepared for the Leslie Harris Centre and Municipalities NL. https://research.library.mun.ca/1966/3/Understanding_Regional_Governance.pdf

Webb, J. A. (1998). Confederation, conspiracy and choice: A discussion. *Newfoundland and Labrador Studies*, *14*(2), 169–187.

# 5 Leadership and participation engendering new perspectives for community development


**Abstract**

*This chapter discusses the fourth and last infrastructure relevant for the rethinking of community futures and assets: leadership. Here we consider in particular distributed leadership, referring to the collaborative work of both formal and informal leaders in the collective decision-making process. Speaking of leadership is thus speaking of participation.*


Governance cannot function without leaders. It can never rely entirely on the routines of a well-designed system of actors and institutions. In communities facing tangible obstacles and the need to confront often deeply entrenched narratives and limitations in local power, leadership is especially important. In communities with lasting legacies of tough or traumatic histories or internal divisions, it can be difficult to engage in self-reflection, while there tends to be mistrust of higher-level governments and suspicion of outsiders. It can be challenging to overcome these obstacles and come to a new understanding of the situation and what might be a strength or asset. Yet, in the management literature, in popular discourse, as well as in politics, stories abound that create unrealistic expectations (Obolensky, 2017; Spoelstra et al., 2016; Storey et al., 2016). Too much is expected from leadership and, consequently, too little of other community members who may not see themselves in a leadership role. We focus on distributed leadership rather than looking to individual heroes, as this taps more fully into the potential of the community, and for the simple reason that community leadership is, in practice, nearly always distributed. Hero worship, however, is an all too common tendency in times of stress and in a context where external, centralized control has dominated.

## What is leadership?

As we have noted, in the context of community development, leadership is best understood as distributed leadership, with each place and region revealing a different combination of individuals, organizations, and communities

DOI: 10.4324/9781003620297-5
This chapter has been made available under a CC-BY-NC-ND 4.0 license.

playing a role. The ability to bring in new ideas, translate them into relevant interpretations of local possibilities, bring together people in a new way, mobilize resources, and redefine problems and solutions can be recognized as leadership qualities. Rethinking assets, persuading others to do so, and lifting barriers for those formerly excluded or impeded to rethink what could count as an asset – all this requires leadership.

These functions of leadership can exist outside government and even outside governance. When it comes to reflecting on assets and underlying narratives, leadership can connect people, create trust and a safe, open environment where old grievances and new ideas can both come up and where values can be both defended and questioned without formal or informal repercussions. Leadership, moreover, can navigate the divide between insiders and outsiders and their varying ideas on problems and opportunities, as well as community identity and assets (Caine et al., 2007).

In a context where change is needed, leadership will have to embark on community self-analysis leading into a phase of strategizing (and later implementation). An initial understanding of the situation by leadership can trigger more participatory forms of self-analysis. Nothing might happen without an initial push. Once self-analysis is under way, encouraging communities to look at themselves differently, their past and desired future, will be challenging, and leadership most likely has a role. Later, crafting strategy, coming to a shared new narrative and linking it systematically to policy tools will require different things from leadership. Finally, implementation will be more than pushing a button. The process will entail continuous observation, of the relation between governance, community, and the ecological environment, and tactics by leadership to keep the strategy on track (or modify it as needed).

When communities are tightly coupled to their natural environment, and when serious questions arise about the viability of the community, leadership comes under enormous pressure (Brosius et al., 2005; Van Assche et al., 2016). Sharpening and diversifying the observation of the relations between community and environment will be essential. Governance then must rearrange itself, providing space to discuss, assess, and create long-term perspectives that build on existing stories and values in a critical and discerning manner. In this way, long-term perspectives are ensured to remain alive in governance, and monitoring the observation of system-environment relations becomes a key task for leadership when sustainability becomes a topic. We speak of second-order observation, or the observation of observations, as gaining in importance, when the community has difficulties in relating to its environment (Van Assche et al., 2024).

## Myths on leadership

Leaders do not have to be highly qualified experts in anything. They may be experts about their community. A common myth is that leaders must be the best at something: the best at speechifying, the best at distilling the essence from a situation, or the most proficient in a particular kind of *knowledge* (Alvesson & Spicer, 2010; Van Assche et al., 2016). Instead, we argue that leadership in a community context requires balancing different types of knowledge *in governance* (not within one person): local knowledge of different sorts, traditional knowledge, expert knowledges stemming from different disciplines, and expertise in administration and its various tasks.

It is helpful for leaders to see how certain forms of knowledge are linked to dominant narratives, even with themselves. Despite the inclination to see this as commonsense, leadership should not focus solely on finances, and see obstacles for development primarily in financial terms. Certainly, money is helpful, but a financial focus is not the best way to rethink assets and strengths and develop long-term perspectives. Money is a means to an end in community development, and leaders must be able to manage organizational finances for the public good. If financial troubles occur, they must be addressed but this is not the same as crafting a strategy (Carter et al., 2008). Ideally, addressing financial woes goes together with strategizing, where those woes ought to be reinterpreted – what are the assumptions in interpreting the problem as financial? What other problem also exist? And what assets might be brought to bear?

Perfect institutional arrangements or governance systems are not required for leaders to make a difference. Institutional reform can be inspired by leadership, and might make leadership more effective and less dependent on personalities. Nonetheless, we can argue that where governance capacity is limited, leadership is even more necessary to make governance function and ideally develop capacity. Just like perfect governance systems do not exist, neither do perfect leaders. We suggest that leaders should not presume or strive for perfect institutions in terms of approximating an ideal design but rather strive for improvement and adaptation, being aware of the types of learning present in governance and community and finding ways to develop or reconfigure them.

Leaders are neither heroes nor saints or prophets, and they do not need to be. They are not required to demonstrate extraordinary courage or virtue. Moral superiority is not part of the job description; neither is the possession of a crystal ball to see into the future. If leaders are expected to be super-human, they will ultimately either alienate or disappoint those they are leading. If excessive hope is invested in individuals, a thorough reflection on the structure and process of governance is needed (Einola & Alvesson, 2021).

## Stories and leadership

Leadership can help communities reflect on its stories and how they are structured, as well as helping craft new stories. Yet leadership does not exist outside the world of stories itself. Communities do not only *produce* their own type of leadership; they also *require* their own type of leadership. They tell *particular stories* about what leadership is and should be (Czarniawska & Rhodes, 2006).

What could bring existing leadership and desired leadership closer together when change is needed is replacing and then redefining leadership, in a combined process of governance reform and cultural change. This is theoretically possible, but unlikely to be easy. Another approach is to encourage self-reflection within the community and the governance system. This too is challenging, yet often much more realistic, and in line with the approach of this book. A third option is to strengthen outsider perspectives, temporarily or more structurally, by bringing in advisors, supporting internal outsiders, or learning about other perspectives – as in Wylezol's story below about visiting geologists and their contribution to new stories and to local leadership. Finally, one can find new avenues for participation in local governance that aim to diversify perspectives, for example by bringing in both outsiders and/or formerly silenced insiders.

If communities are tangled up in their own stories, leadership can assist in the untangling. In the next chapter, we present a method of self-analysis, revolving around the four infrastructures identified, where leadership, in all likelihood will be implicated, possibly taking the initiative, maybe helping to structure the process. Also, outside such more ambitious efforts, and in diverse and everyday settings, leadership can gently encourage citizens to open up their perspectives on past and future, to question assumptions, and to give alternative stories a chance. Sometimes, the impetus can be critical, interrogating previous leadership, their closed narratives; elsewhere, the power of a new story can make one forget the old ones and their limitations. In all cases, it is useful for leadership to reflect on the stories that keep obsolete understandings of assets, and possibly of leadership and governance, in place, and, at an appropriate time, to share this reflection with the community.

---

### Leaders as builders, translators, and the uncovering of unseen assets

### Western Newfoundland – pot of gold at the end of the Appalachian rainbow

Paul Wylezol

When people think of the "Rock," as Newfoundland is fondly known, they often think of picturesque fishing villages, quaint dialects, offshore oil, but rarely do they think of…rocks. However, that is slowly changing, as this *micro-continent* on the east coast of North America is at the geologic and academic center of what became known as the Iapetus Ocean.

In the 1960s, geoscientists and geologists such as Newfoundland's Harold "Hank" Williams were contemplating a theory of continental drift known as plate tectonics whereby lithospheric plates that compose earth's outer shell move across a semi-fluid upper mantle. Hank's work in Western Newfoundland, along with that of other geologists like Robert Stevens, led to a more comprehensive understanding of the Appalachians Mountains, which in turn culminated in the compilation of his renowned *Tectonic Lithofacies Map of the Appalachian Orogen*, published in 1978. Their work also resulted in the creation of Newfoundland and Labrador's Gros Morne National Park and its designation as a UNESCO World Heritage Site. Indirectly, Hank and other geologists' extension of the Appalachian Mountains into Canada's Maritimes and Newfoundland led to the creation of the International Appalachian Trail (IAT) in 1994.

The Appalachians' recognition as the sibling of the Caledonian Mountains led to the IAT's extension to Western Europe in 2009. In return, the Global Geopark Network, established in Europe in 1998 and now a UNESCO designation, has been extended to eastern Canada, with Geoparks in New Brunswick, Gaspe Peninsula of Quebec, Nova Scotia, and Newfoundland (i.e., the Bonavista Peninsula's Discovery Geopark), with a new Geopark being developed across the spectacular Bay of Islands Ophiolite and Humber Arm Allochthon in Western Newfoundland. This may be aptly described – and confirmed by Hank's Tectonic Lithofacies Map – as the pot of gold at the end of the Appalachian rainbow! (Figure 5.1).

*Figure 5.1* Illustration of the "pot of gold" or Tablelands in Gros Morne National Park, Western Newfoundland. Credit: Paul Wylezol.

## Local leadership and learning from Newfoundland and Labrador

Where communities are under pressure, where assets and futures require rethinking while such rethinking is hampered by issues with the infrastructures discussed, leadership might have to take on specific roles.

The first role we distinguish is that of *leaders as builders*. This role refers to the constant task of picking up the pieces aftershock, or after histories of neglect and trauma, trying to put them together and find the elements that are needed to rebuild communities and livelihoods. Leadership in this role is thinking constantly about *what is missing*, *what is needed to maintain or improve what is understood as good governance and quality of life*, and what pieces are in place/available to be repurposed.

Often, there is a need to build institutional capacity or the ability to make collective decisions and translate them into collective action. Sometimes this is a matter of small additions or modification, while in other cases it means starting from scratch or overhauling the governance system. The obstacles and tasks for leadership will hinge on the community's support and, of course, what needs to be built. The benefits of this new infrastructure, therefore, must be carefully explained to all. Building can rarely be separated completely from the other roles of leadership, and in this sense "explaining" the benefits often involves persuading and resonating with the values and stories that are important to the people (Sveningsson et al., 2012).

Hence, we identify a second role of the *leader as broker*. Brokers build connections that might make things possible because they have a good understanding of power relations and the levers and buttons that can alter them. A broker can help resolve conflict but also establish external relations or internal coalitions. Builders often find themselves having to choose between quality and cost, or between esthetics and functionality, which leaders must also do in their role as builders. Leaders need allies just like the community itself needs allies – those who are willing to envision an alternative future based around what the leader is building. Thus, leaders play a strong role as brokers.

Brokering, however, does not come with clear and simple instructions, while also carrying risks. Brokers can manipulate either of the parties they are connecting or amass power that is not easily observed. In other words, while solving problems, they can also undermine checks and balances, as they can create new networks outside of democratic control (Kamp, 2018). Simple instructions are not possible, because brokers deal with different things that could be connected in different strategic and cultural situations.

Brokers can use many skills, tactics, and strategies to build connections and coalitions. They may connect not only individuals, groups, organizations, levels of government, and public or private actors but also ideas, stories, and perspectives on the world. Stories might be unstable and might also be invested with different feelings for different parties, sometimes latently. This can apply to the stories brought together by the broker, and to new stories crafted by the broker, with the hope of redefining a common ground. Stories, moreover, form unexpected connections with other stories or lose persuasive power because one of their ideas, metaphors, or other connected stories may lose that power. Mayor John Norman tells the story below of Bonavista and how multiple levels of government, citizens new and old, and a local Chamber of Commerce came together to envision and create a new future for the community based on a re-examination of local assets.

---

**Developing restorative capacity**

**A community's self-reflection journey, the re-examination of its assets, and the different roles of local leaders**

***A conversation with John Norman, mayor of Bonavista***

As roads were being paved in the region of the Bonavista Peninsula, the population was falling and the economy failing. People felt it was time to seriously think about the future. They were concerned about the loss of community vitality but were reluctant to create a tourism Disneyland that would come to life in the summer and die back in the winter. While many communities focus on what they don't have and try to build new, people in Bonavista asked themselves, "what do we have that can help us grow and stabilize the community?".

The municipality embarked on an asset inventory, revealing that by far the most numerous assets were old buildings scattered through a unique and beautiful landscape. Rather than seeing this as a liability, remnants of the once vibrant community that lived there, the municipality saw an opportunity. With a shrinking economy, building new was not a realistic option, so creating spaces for living and working would have to be done with what was there. Efforts were made to lobby for provincial and federal grants to restore and revitalize these properties. What would result from those efforts was initially not entirely predictable, but the opportunity was clear; restoration meant revitalization, and that meant possibilities for diversification.

Nowadays, over $40 million in business is done in tourism alone in Bonavista over six months of the year. Although fishing was and still is an important activity for the community, local livelihoods and identities have diversified beyond tourism. The town's economy is a creative economy, with artist and architecture studios, business incubators, a wallpaper studio, restaurants, coffee shops, ice cream parlors, etc. The town has become a charming destination not just to visit but also to live. People coming from different places of the world find in Bonavista a space to belong. The restored buildings are part of a creative landscape that has uniquely been cultivated by the community and that has resulted in a new economy. Businesses are moving in from different places in Canada, and businesses are being created. Sometimes this stems directly from the focus on heritage, as with the local business specializing in the restoration and production of heritage windows.

Opening space for the community has been the core of change in Bonavista. It has now created an environment, an entrepreneurial ecosystem, a creative ecosystem that is generating and attracting social enterprises, new businesses, and new leadership in the town. Young people who have historical connections but also newcomers in their twenties and thirties are getting involved in the community; they are volunteering with the local Chamber of Commerce, the local theater that has been reopened, and launching new festivals. For a community, it is very important to understand the role of attracting these individuals not just as taxpayers but to lead the community in different ways.

Bonavista is committed to the development of community, not simply business. When the focus is community, business will come. The opportunities that came out of those original exercises – identifying assets and thinking of how to use them – allowed the community to renovate. First it looked as if we were dealing with just buildings, old buildings, heritage maybe, but along came a renovation, a rejuvenation of our community identity.

---

In places where outsider perspectives are not accepted and (sometimes over-stable) community perspectives on life and the future, leaders in their role as brokers will need to find common ground between insiders and outsiders. This could take the form of identifying shared interests, values, and care for the community (Elias & Scotson, 1994). Stories on each side can be interpreted slightly differently to reveal possible overlaps. It means that leaders need to know both the community and its larger environment, navigating both landscapes and explaining things differently in two or more directions.

The previous considerations bring us to our third role of leadership: *leading as translating.* In the activities of the leader as a broker, such translation is implied as a continuous activity and concern, but it also comes in handy elsewhere. Some leaders might not act as builders of new grand narratives by shedding new light on community assets, nor build new institutions or institutional capacity. Rather, they bring in and translate ideas, perspectives, and governance tools. Leaders have access to different circles, places, ideas, tools, and histories that are not easily accessible or understandable for other community members or governance actors (Kamp, 2018; Sveningsson et al., 2012), while their power of persuasion often rests on the ability to translate ideas and stories in many different ways.

Finally, we come to our fourth role, *leaders as trusted listeners and advisors*, who rely on a combination of observation and listening skills. In this role, leaders take on a role similar to a therapist. They do not immediately take the initiative to build, connect, or translate, but instead focus on listening, understanding the sentiments and attachments in the community, and ultimately helping the community better understand itself (Brown & Augusta-Scott, 2006a; Karbelnig, 2023). Such leaders might hold up a mirror and gently question whether current practices and beliefs should change or stay the same.

Small groups of key actors, often *de facto* leadership itself, can open themselves up to new approaches when everything else is felt to fail, and when the group believes itself to be a catalyst in broader transformation. Larger groups can come together directly through open meetings, visioning sessions or strategy processes or discussions, and indirectly via local media. The moments, places, and participants must be chosen carefully, recognizing and facilitating the correct conditions for guided self-questioning rather than attempting to lure or force self-questioning. Under such conditions, an openness for change can develop into new interpretations of self and community (Van Assche & Gruezmacher, 2025).

Carefully chosen stories can make a difference, especially when it comes to community pride, identification with traditions, the need for certainty and fixed identities in governance, and the acceptance of critique from either insiders or outsiders (Brenneis, 1988; Brown & Augusta-Scott, 2006). If community members are already upset with leadership, confronting them might be counterproductive. Similarly, leadership portraying itself as all-knowing or assuming they possess an ultimate truth not accessible by the community is unlikely to be well received. On the other hand, if leadership has gained deep respect and trust and the community believes leadership has their collective interest at heart, they are in a better position to gently and cautiously shift perspectives of the community on itself, so that new forms of self-reflection become possible and hopefully open the door for new futures (Alvesson & Einola 2019).

---

## From film and the Fogo process to community fora: The importance of contextualized arenas for public discourse

Fred Campbell

Newfoundland has been recognized globally since the 1960s for using media as a tool for community development. The Fogo process, which took off in 1967, was not the end of the development (see also Newhook above). In 1979, Memorial University technicians started taking portable televisions to communities across the island – many of which hadn't have any access to local television before. About 30 television transmitter projects developed from 1979 to 1989. My own experience began in 1989 when the Extension Service hired me to design and train volunteers to produce community television on locally owned cable systems, together with field workers.

The Fogo process had concentrated on film and video, but other media such as popular theater or puppetry have been utilized in the same way in Newfoundland and Labrador. A range of media have been used in processes to reflect, enhance participation, build consensus, and work toward community and structural change. In yet another example, in 1984, Extension field worker Neil Tilley worked with the St. Fintans Area Development Association on a six-month planning process in the Bay St. George South region. This included night events that featured a moderated panel discussion with a speaker phone so local residents could call in to ask questions and offer opinions on issues of concern. This process would eventually become the *community forum*, in my view one of MUN Extension's greatest gifts to participatory communications.

The purpose of a community forum, as we now define it, is to promote cultural celebration and the public discussion of local issues. The forum was typically formatted in three parts over three days:

Day 1 – Celebrate the past: This not only initiates cohesion among the participants but also makes people feel good about themselves and their communities and more ready to speak out constructively.

Day 2 – Discuss issues: Typically, the panel consists of local people with a moderator. Visiting "experts" and others speak from the audience. Call-outs were used as well as moderated call-ins.

Day 3 – Action plans for the future: People decide what can be done about the issues and who will do what.

In 1993, the Community Education Network (formerly the Port au Port Community Education Network) held a series of public community television events on the Port au Port Peninsula drawing from the community forum experience. Participatory communications continued over the next decade through projects such as Communication for Survival/Communiquer Pour Survivre (participatory rural dialogue in eight communities); Talking About the Zone (public consultation forums using community television); Youth, the Environment and the Economy conferences (mixed media including community television and interactive websites); and Sharing Our Future/Partager Notre Avenir (including local facilitators working with teams of youth volunteers, community television and radio special events, reaching out to expatriates with email, interactive websites, and webcasting).

Community radio is recognized globally as an important tool for facilitating communication within and between rural communities, now often accomplished through webcasts. But solving technical problems is only one thread of implementing successful community fora, and of participatory communications for rural development and sustainability more generally. The lengthy incubation process (six months of intensive stakeholder dialogue recommended) and production costs mean a group of understanding partner/champion(s) is essential along with key elements that include a production team (including youth) and planning committee, musicians, panels, and a participating audience as well as effective promotion (Figure 5.2).

## Conflict, leadership and assets

The rethinking of assets and the reconfiguration of futures in governance is at times impeded by conflict or polarization, which often increases distrust and hardens the difficulties encountered in coming to a shared vision and a new understanding of what might be of value in the future (Andrusz et al., 1996; Chong, 2011). If this is the case, governance is bound to be challenging. Taking collectively binding decisions will be difficult when there might not be much of a collective anymore. Self-analysis, visioning, and strategizing will be more problematic and implementation less likely. Leadership can

*Figure 5.2* Community forum in Lourdes 1993 sponsored by the Port au Port Community Education Initiative, later the Community Education Network (CEN), with a moderated panel discussion, community participation, equipment operated by local youth, and locally broadcast by a low-power television transmitter. Credit: Fred Campbell.

be instrumental in overcoming these issues but, in such cases, it is likely that leadership is also part of the problem.

Where leadership is blind to latent conflicts or polarization or if it is unaware of how their actions can exacerbate them, even the best-crafted strategy and the most honest self-analysis can be deliberately misinterpreted, creating further rifts (Aasen, 2017; Warner, 2019). If this happens in a democratic fashion, an ensuing change of guard is legitimate, but coming to a new and legitimate community strategy will prove much harder in the future. Leadership can also be intentionally blind to latent conflicts, deliberately ignoring particular actors, stories, local knowledge, or maintaining broader patterns of exclusion. Problems of exclusion might exist where neither participation nor representation processes work to link certain groups or voices to governance, which can fuel discontent and, in due time, conflict and a refusal to participate in the future (Fung, 2006).

Conflict can be productive, yet it must come to the surface and be managed for there to be hope for some form of collective visioning and reconsideration of assets (Del Lucchese, 2009). Thus, leadership must become aware of latent conflicts and hidden divisions, while distinguishing between productive and unproductive conflicts that undermine governance and trust in institutions. Traditions of managing or avoiding conflict can be both helpful and problematic, as an aversion to conflict can lead to blindness to conflict, weaken critical capacity, or lead to avoidance of tensions that can then escalate into larger conflicts and related community challenges. In contrast, fondness for direct confrontation risks disrupting social relations and governance routines, which can easily leave memories that become problematic by themselves. Creating a culture of direct communication in any organization or governance system can make it easier for individuals to raise issues that require re-evaluating a routine or decision-making process.

Regarding *external conflict*, higher-level governments have tools available to defuse tensions and rivalries between communities and address problems between themselves and lower-level governments. But of course, if senior governments are in conflict with a given community or group of communities, these tools are less useful due to the biased viewpoint that actors in a given department or agency may have toward the communit(ies) in question. Second-order observation may provide advantages such as reduced pressure to identify with one perspective or less acute anxiety about the future. At the same time, the power imbalance between senior government agencies and communities implies that the former's perspectives on future threats and opportunities can be imposed onto communities under the guise of concern for the public interest or the broader polity's welfare.

Conflicts can revolve around different interpretations of assets or the question of whether something is an asset at all. These conflicts render the collective rethinking of assets and futures more difficult. On the positive side, a new asset can emerge out of a conflict if the tension produces a new perspective on the environment or the future, which in turn leads to the discovery of an asset. Leadership in conflict situations thus becomes more important in the different roles distinguished above, yet leaders must be even more careful. Leadership must do more, but every move will be met with more suspicion (Kernberg, 1998). Therapy might be more needed in situations of conflict but where trust in leadership is feeble, the prospects for collective reflection of a therapeutic nature are weakened.

---

**Hydroelectricity and community engagement in energy development: Muskrat Falls as an asset?**

Mark C.J. Stoddart and Karin Buhmann

Muskrat Falls is a hydroelectric megaproject located on the lower Churchill River in Labrador. Under the Conservative Government of Danny Williams (2003–2010), the project was conceptualized and approved within a broader political and public narrative of the Churchill River as an asset that should be used to advance economic development for the province as a whole, and therefore coming under the decision-making purview of the provincial government. It was also articulated as an optimistic narrative of balancing the perceived losses vis-à-vis the province of Quebec over the earlier hydroelectric development of the Upper Churchill Falls project (House, 2021). This narrative was a key factor in driving the project forward despite red flags raised by the impact assessment process, as well as by public critics of the project who questioned its economic feasibility.

As the project was pursued in a manner that seemed pre-determined (LeBlanc, 2020), various tensions and counter-narratives emerged that highlighted issues with public participation in energy development on the Churchill River. The impact assessment and other review processes set expectations for public participation and engagement before the project moved forward. Early stages of engagement focused primarily on constructing a community benefits agreement with the Innu Nation. However, tensions emerged around those who were defined as affected communities beyond the Innu Nation, and who were included or excluded.

Concerns were raised by other downstream communities, including Nunatsiavut and NunatuKavut governments and community members, and included possible environmental health risks of increased methylmercury concentrations in downstream fish and wildlife. Concerns also surfaced about downstream flooding due to the slope stability of the North Spur part of the dam and changes to the river's flow regime and impacts of the transmission line.

The project continued to go over budget and behind schedule. This largely proved public critics of the project correct and led to a sense of financial crisis for the province. It ultimately led to intervention from the Federal Government to help reduce the debt load from the project, which would otherwise be passed on to ratepayers. This eroded public trust in the project proponent (Nalcor) and the provincial government and reinforced the sense that Muskrat Falls was going to move ahead while community concerns were downplayed and minimized.

How might we rethink the narrative of the Muskrat Falls and the Churchill River as an asset? While the Churchill River can be seen as an energy and economic asset, it is also an asset for land-based relationships of hunting and fishing, contributing to food security, a sense of connection with the landscape, and recreation. Without prescribing our own definition or new narrative, we suggest several practices for more meaningful public engagement and consultation that can lead to new narratives of the Churchill River that are more reflective of the experiences and interests of diverse local communities.

Proponents and decision-makers should spend more time "on the ground" to understand local needs, interests, concerns, and social dynamics within and across different communities. Engaging in consultation earlier in development processes, including at the visioning stages, is also a way for affected communities and individuals to help shape the narratives of how the river is valued and how it should (or should not) be treated as an asset for development. In the case of Indigenous communities, there is also a legal duty to consult. When projects do move forward, it is also important to ensure they are used to invest in community infrastructure and provide a legacy that is viewed as positive from the perspective of those at risk or affected. Finally, for rightsholder and stakeholder engagement to be experienced as meaningful, it is important to embody transparency, reflexivity, and openness to learning (and change) among project proponents and decision-makers.

Muskrat Falls has come into operation and is producing power, though technical and infrastructure issues continue. It is vital to learn from Muskrat Falls. Treating the Churchill River as a provincial economic and energy asset in narratives that promoted and legitimated Muskrat Falls failed to take into account the tensions and alternative narratives and values that emerged as the project was pushed through. Issues of trust, transparency, and meaningful engagement are central to revisiting and recreating narratives about the Churchill River to reflect local interests and needs.

## Concluding

In this chapter, we discussed leadership as an infrastructure for rethinking assets and building new futures based on those assets. Leadership is best understood in this context as distributed, and the skills needed will differ in each community (Jones, 2014; Van Assche et al., 2024). For communities facing the issues discussed in this book, issues exemplified by many communities in Newfoundland and Labrador, we argue that helpful leadership roles include those of translator, broker, advisor, and builder. While building is necessary in the long term, for most communities in Newfoundland and Labrador – as well as many small, remote, resource-dependent communities with limited institutional capacity anywhere – building might become a possibility only after phases of translating, brokering, and possibly therapy. In many cases, this may be possible after episodes of conflict management. In relation to managing conflict, one possibility is leadership change. Another is fostering leadership development. Maintaining diversity in governance perspectives helps to widen the pipeline for leadership as it is useful in introducing flexibility and creativity in the framing of assets. The different infrastructures for rethinking assets and futures thus support each other, a point we will return to in the book's conclusions.

In crafting strategy, it is fundamental for leaders to recognize that the process will require collective effort and might take many forms, but that whatever is defined as a strategy needs to integrate not only policies but also the desires and aspirations of people (Golsorkhi et al., 2010; Mintzberg & Lampel, 1999). Hence, the need for leadership to craft a strategy sensitive to collective values and understandings and the need to observe continuously the effects of strategy on the community and on governance itself (Auvinen et al., 2013). If people are not able to see themselves beyond their past or learn from the experiences of other communities, and if they cannot see what in their daily life is the result of community strategy, its legitimacy and efficacy will be undermined.

## References

Aasen, M. (2017). The polarization of public concern about climate change in Norway. *Climate Policy*, *17*(2), 213–230.

Alvesson, M., & Einola, K. (2019). Warning for excessive positivity: Authentic leadership and other traps in leadership studies. *The Leadership Quarterly*, *30*(4), 383–395.

Alvesson, M., & Spicer, A. (2010). *Metaphors we lead by: Understanding leadership in the real world*. Routledge.

Andrusz, G., Harloe, M., & Szelenyi, I. (1996). *Cities after socialism: Urban and regional change and conflict in post-socialist societies*. Blackwell.

Auvinen, T., Aaltio, I., & Blomqvist, K. (2013). Constructing leadership by storytelling – The meaning of trust and narratives. *Leadership & Organization Development Journal*, *34*(6), 496–514.

Brenneis, D. (1988). Telling troubles: Narrative, conflict and experience. *Anthropological Linguistics*, *30*, 279–291.

Brosius, P., Tsing, A. L., & Zerner, C. (2005). *Communities and conservation: Histories and politics of community-based natural resource management*. Rowman Altamira.

Brown, C., & Augusta-Scott, T. (2006). *Narrative therapy: Making meaning, making lives*. SAGE Publications.

Caine, K. J., Salomons, M. J., & Simmons, D. (2007). Partnerships for social change in the Canadian North: Revisiting the insider–outsider dialectic. *Development and Change*, *38*(3), 447–471.

Carter, C., Clegg, S. R., & Kornberger, M. (2008). *A very short, fairly interesting and reasonably cheap book about studying strategy*. SAGE.

Chong, D. (2011). Coordination and conflict. In *Rational lives: Norms and values in politics and society* (pp. 76–115). University of Chicago Press.

Czarniawska, B., & Rhodes, C. (2006). Strong plots: Popular culture in management practice and theory. In *Management education and humanities* edited by Pasquale Gagliardi and Barbara Czarniawska, (pp. 195–120). Edward Elgar Publishing.

Del Lucchese, F. (2009). Crisis and power: Economics, politics and conflict in Machiavelli's political thought. *History of Political Thought*, *30*(1), 75–96.

Einola, K., & Alvesson, M. (2021). When 'good' leadership backfires: Dynamics of the leader/follower relation. *Organization Studies*, *42*(6), 845–865.

Elias, N., & Scotson, J. L. (1994). *The established and the outsiders: A sociological enquiry into community problems*. Sage.

Fung, A. (2006). Varieties of participation in complex governance. *Public Administration Review*, *66*(s1), 66–75.

Golsorkhi, D., Rouleau, L., Seidl, D., & Vaara, E. (2010). *Cambridge handbook of strategy as practice*. Cambridge University Press.

House, D. (2021). The Danny Williams Government, 2003–2010: Masters in our own house? *The Journal of Newfoundland and Labrador Studies*, *36*(1), 1719–1726.

Jones, S. (2014). Distributed leadership: A critical analysis. *Leadership*, *10*(2), 129–141.

Kamp, A. (2018). Assembling the actors: Exploring the challenges of 'system leadership' in education through Actor-Network Theory. *Journal of Education Policy*, *33*(6), 778–792.

Karbelnig, A. M. (2023). *Lover, exorcist, critic: Understanding depth psychotherapy*. Phoenix Publishing House.

Kernberg, O. F. (1998). *Ideology, conflict, and leadership in groups and organizations*. Yale University Press.

LeBlanc, R. D. (2020). *Muskrat falls: A misguided project*. https://www.muskratfallsinquiry.ca/final-report/

Mintzberg, H., & Lampel, J. B. (1999). Reflecting on the strategy process. *MIT Sloan Management Review, 40(3)*. https://sloanreview.mit.edu/article/reflecting-on-the-strategy-process/

Obolensky, N. (2017). *Complex adaptive leadership: Embracing paradox and uncertainty*. Routledge.

Spoelstra, S., Butler, N., & Delaney, H. (2016). Never let an academic crisis go to waste: Leadership Studies in the wake of journal retractions. *Leadership*, *12*(4), 383–397.

Storey, J., Hartley, J., Denis, J.- L., & 't Hart, P. (Eds.). (2016). *The Routledge companion to leadership*. Routledge.

Sveningsson, S., Alvehus, J., & Alvesson, M. (2012). Managerial leadership: Identities, processes, and interactions. In Stefan Tengblad (Ed.), *The work of managers: Towards a practice theory of management* (pp. 69–86). Oxford.

Van Assche, K., Beunen, R., & Duineveld, M. (2016). Citizens, leaders and the common good in a world of necessity and scarcity: Machiavelli's lessons for community-based natural resource management. *Ethics, Policy & Environment*, *19*(1), 19–36.

Van Assche, K., Beunen, R., & Gruezmacher, M. (2024). *Strategy for sustainability transitions: Governance, community and environment*. Edward Elgar Publishing.

Van Assche, K., & Gruezmacher, M. (2025). Psychoanalysis and Governance: Discourse and Decisions, Identities and Futures. Taylor & Francis.

Warner, B. P. (2019). Explaining political polarization in environmental governance using narrative analysis. *Ecology and Society*, *24*(3). https://www.jstor.org/stable/26796972

# 6 An approach to reflection and re-evaluation of community assets and futures

**Abstract**

*In this chapter, we discuss an approach that can be helpful for communities to reflect on and reconsider their existing assets, discover new community strengths and opportunities, and construct strategies built around those assets to enhance community well-being. What is presented here can be understood as an umbrella method, or methodology, which can assist communities to assess why it may be difficult to rethink what is of value. Like any good umbrella, something should fit under it – in this case a series of methods to support steps in the process of self-analysis.*

We distinguished in previous chapters several infrastructures and conditions that are helpful for rethinking assets and building strategy. After a community assesses the state of these infrastructures (which can be present, absent, functional, or dysfunctional to different degrees), it will be better positioned to rethink assets or strengths. Mapping, understanding, and improving these infrastructures is a key method for finding new flexibility in thinking and acting as a community. Here we offer an umbrella method that must be consciously specified and adapted by the community itself, which will yield outcomes unique to each community.

We emphasize that, as part of the assessment of these infrastructures, including actually rethinking assets, and in the context of strategizing, a wealth of methods can be deployed to examine a community's specific aspects and its context in more detail. There is a tendency in much academic literature and practitioner resources to seek a concrete conclusion or tangible outcome through stepwise methods. Attractive as this might be, for the task of rethinking assets suggesting fixed methods for strategy, or even for the rethinking of assets would be misleading. By skipping this important self-assessment of what hampers and enables such rethinking, communities run the risk of overlooking key factors influencing community futures.

DOI: 10.4324/9781003620297-6
This chapter has been made available under a CC-BY-NC-ND 4.0 license.

## Preliminary mapping

We believe that for understanding all four infrastructures, it is useful to have a basic understanding of the history of the community and its governance system. Tracing its evolution is valuable for understanding the present, even where governance was largely informal until recently.

Second, an initial mapping of assets and key narratives is helpful. This includes stories about the past, present, and future, while relations with the environment are brought up but not analyzed in detail. Only stories relevant for the community and its selection of assets are important at this stage (Alevizou et al., 2016; Mintzberg, 2017). Mapping narratives and assets is helpful for all four infrastructures. Mapping governance, for example, helps us understand where institutional reform or building might be needed and where learning and non-learning occur. This renewed reflection brings to light how things are done in the community. This means that productive tensions might arise in discussing various versions or interpretations of roles and responsibilities. An initial mapping of governance will only work if people are willing to ask sincere questions about how things actually work, not simply referring to official rules and regulations describing how things are supposed to work.

The mapping of assets, or strengths, and narratives ought to be linked from the beginning, uncovering the assumptions, values, and ultimately narratives to which various assets connect. Conversely, one can inquire about key stories and, from there, discuss which assets might be enshrined by them. Mapping stories is limited by practicalities, but the exercise in and of itself has no limits. The mapping of assets to narratives, and narratives to assets, makes it abundantly clear that we are in no way aiming for a fixed map, but rather for a starting point for further analysis.

Figure 6.1 outlines the general structure of the approach we propose, starting from preliminary mapping and then moving to a more detailed investigation of each of the infrastructures. Some insights offered into the strengths and

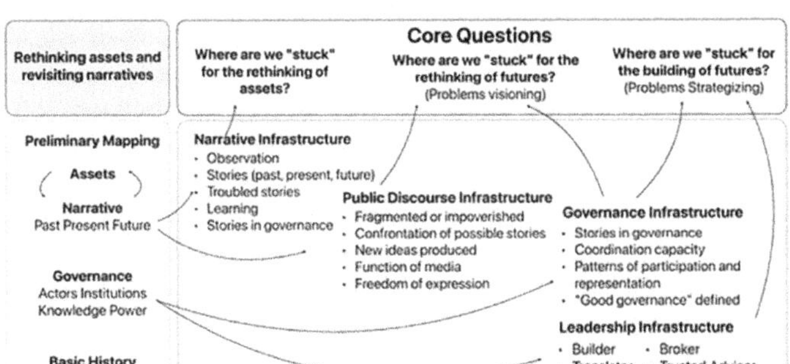


*Figure 6.1* Assessing issues in the main infrastructures necessary for rethinking assets and futures.

*Table 6.1* First mapping guiding questions to be asked sincerely and openly

| |
|---|
| Which periods in the history of our community can be discerned, which actors were involved, which coordination tools or institutions were used? Which different arenas? |
| How did power relations, patterns of inclusion and exclusion *actually* shift? |
| Which forms of knowledge, expertise, stories played an important role? |
| What was the role of administration and its departments? |
| How was the environment seen and used? |
| Did long-term perspectives and strategies play a role? |
| How were the relations to external actors both private and public? |

weaknesses of those infrastructures are beneficial beyond the identification of difficulties in rethinking assets: they can also support strategy. We devote the rest of this chapter to the multiple benefits of the method. As communities engage in self-assessment of each of the infrastructures, this process will naturally bring up several methods that could be deployed to support that effort. The Appendix presents a sample of additional methods, some of our own devising, which can play a role in the context of the proposed self-analysis.

Self-analysis is crucial for the types of community we have focused on in this book, in which the prevailing economic and other assets have been threatened; yet this process can also be more difficult in these contexts. Recognizing problems is not easy, nor is acknowledging that assets which once underpinned community livelihoods may have changed in importance. Familiar narratives, values, and identities can be comforting, but may also create trouble when this rethinking is necessary. Even seemingly simple forms of mapping governance, assets, and narratives often carry sensitivities. Guidance will be needed, either by local leadership, empathetic outsiders, or a combination of both. Therefore, it is important to emphasize that it is entirely up to the community to establish the limits and direction of this self-analysis (Mohatt et al., 2014; van der Watt, 2018).

An initial mapping must involve consideration and articulation of a community's own guiding questions, yet Table 6.1 can serve as an orientation with sample questions to structure the mapping of governance. As noted, this must be contextualized, and different questions might be sensitive in different places or difficult to answer for other reasons. The point is to spark reflection on the nature of the community and its governance system and probe how locals understand – and want to understand – their place. Where reflecting on governance proves a difficult starting point, one can also embark on the exercise by asking questions of assets and narratives, choosing to discuss assets first and from there explore narratives, or work the other way around.

Where none of this seems appropriate, one can also consider starting with the arts. While arts-based methods can figure prominently, if one desires so, in the analysis of several of the infrastructures (see also the Appendix) one can imagine situations where what is needed first is a loosening of perspectives, roles, and maybe even power relations. Art, in the broadest sense, can be powerful in shifting the perspective of people and groups, loosening hardened relations, and revealing new qualities in places and people (Lowe, 2000). For example, recall the reflections of children on the Port au Port Peninsula shared in Chapter 1. As with the leadership style based on leaders acting as trusted listeners and advisors – in which arts-based methods may be particularly useful – the conditions must be right. When people feel obliged, alienated, or unsafe to express themselves authentically, art (as in other forms of community reflection) can further entrench positions or even create new rifts.

When a community or its leaders feel ready for further investigation and to really dig into the infrastructures, Figure 6.1 can provide a structure for this process. In the case of a full self-analysis, we recommend going through the four stages and then considering: are we in a better or freer position to rethink assets and futures? We anticipate that, in the process of creating space for rethinking, this rethinking will already have begun. One can also predict that elements of a strategy will begin to appear, making self-analysis immediately valuable for a strategy process. One final introductory note is that people may decide early on that one or more of the infrastructures are most relevant for their community, focusing efforts on those domains. The rest of this chapter will discuss the assessment of each infrastructure, starting with narratives (Figure 6.2a and b).

## Narrative infrastructures

When a knot of narratives exists around an asset that clearly needs rethinking, unraveling that knot can take time. It can stir up conflict, and it might have to be approached indirectly, perhaps by first focusing on seemingly unrelated issues. It can be hard to predict what might happen when alternative perspectives confront us, or the origin of one perspective is unraveled and seen in a different light (Czarniawska & Gagliardi, 2003). It is entirely possible to open up identity narratives or discover or re-articulate core values, which opens the door to a different selection of assets or a new use of existing assets. Opening up these understandings can come from the *observation of observations*, confrontation with different perspectives, and thinking about the history of narratives and practices around a given asset (Kapitan et al., 2011; Steiner, 2012).

It is also an option to start from the future by discussing the desired, possible, and feared futures that may have emerged out of an initial narrative mapping, then working backward to current assets. An opportunity to reflect might stem from feared futures, or a discussion of possible or desired futures. Fear or fantasy might have colored the imagination of the future. When people allow these affects to come to the surface and discuss them, it might support a more open reflection on what they value in the present (Gunder, 2015; Wardle, 2016; Žižek, 2017). Even in the absence of strong emotions, one can consider a desirable future and infer what ought to be present to move in that direction. Such

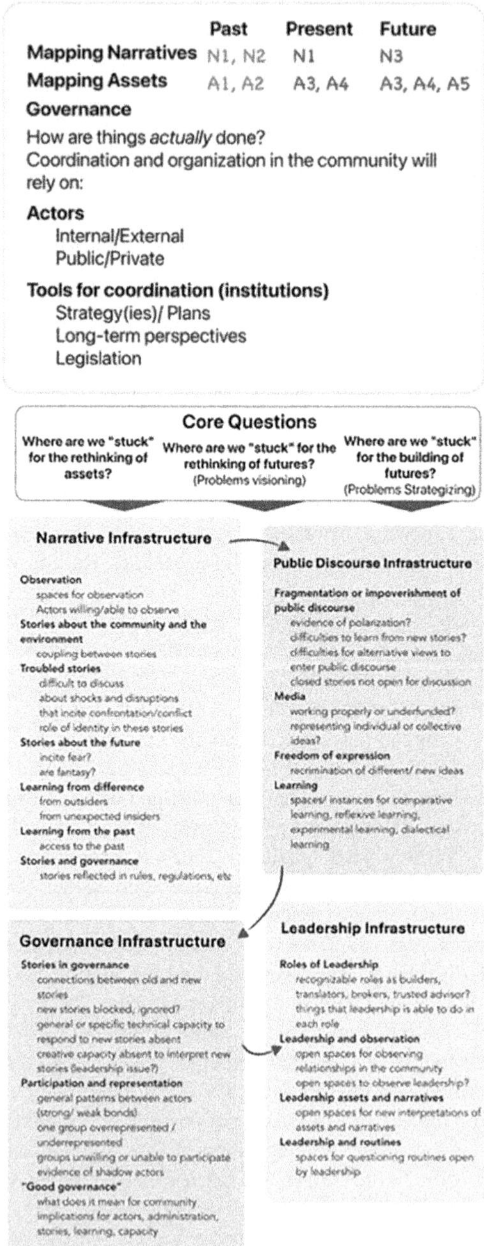


*Figure 6.2* (a) Initial mapping of past, present, and future narratives and assets as well as mapping of governance and its elements. (b) Details and guiding questions in assessing issues in key infrastructures necessary for the rethinking of assets and futures.

an exercise can trigger a rejection of the imagined future, perhaps labeling it as unrealistic. Alternatively, a reflection on current assets might quickly lead to a discovery of assets hidden in plain sight, or opportunities to craft new assets.

We must resist the urge to "fix" narrative infrastructure, since the reconsideration of possible futures is not about identifying the correct narratives but creating *space for discussion and self-reflection* about the community's stories. "Correct" narratives do not exist, as communities define their identity, past, and future in the ways they prefer (Brown & Augusta-Scott, 2006). As with any research instigated by outsiders, there is a risk of seemingly "unbiased" observers assuming that they see what is wrong with a community's prevailing narratives and that they know what narratives would serve the community best, but if community members are not deciding which narratives to question it will be yet another externally imposed intervention (Denscombe, 2024). What is realistic, however, is a self-exploration where communities may find that some tightly held assumptions are bolstering community resilience, while others are keeping practices and ideas in place that are self-limiting or problematic in other ways. Pointing out consequences, checking assumptions, and comparing alternative narratives and practices can play a useful role. An unrealistic future, or a future that is overly stable and narrowly defined, can reveal its problematic assumptions easily, or one might have to trace back the issue to a knot of seemingly unrelated stories to find the root of the issue.

Second-order observation, or the observation of observations, can be practiced in self-analysis if groups can reflect on the way stories are built, circulate in the community, and shape thinking. Different perspectives can be brought in relatively easily. *Learning from difference* is usually easier than cultivating second-order observation and then learning from it (Buchinger, 2007; Hahn et al., 2008). Some methods that are useful for second-order observation also work for introducing differences.

Understanding the past in a different way, unraveling the path that made some stories particularly important, can take place purely through internal discussions but can also be done by bringing in historical experts, dusting off archival material, or conducting new historical studies. It can be cultivated by bringing in outsiders, but also through role-play, community theater, traveling, comparing communities, and other methods (see the annex).

Revisiting the past can be accompanied by interventions in the public domain, such as a public lecture or discussion, a historical exhibition, a documentary, a work of art, or even a community theater production exploring a locally important theme (Kapitan et al., 2011). Local radio, newspapers, or social media can offer opportunities to revisit the past, ideally with an informed and neutral moderator (Gordon, 2012; Howley, 2005; Jeppesen, 2016). From a reinterpreted past, inferences can be made regarding present assets and potential futures. Similar venues can be used for deliberate examination of current strengths and challenges, as well as possible futures, which often link back to a community's past and related narratives, as illustrated by the case of Corner Brook's industrial waterfront below.

## Mapping possible futures and empowering communities: embracing foresights for tourist-friendly development of cruise corridor through engaged learning and partnerships

Umme Kulsum and Kelly Vodden

Integrating innovative approaches that build on the heritage and other assets of a community has become critical for sustainability. Identifying and stewarding community assets is an ongoing, long-term community development process, which can benefit from innovative future-oriented strategies and methodologies in the era of rapid environmental, social, and economic changes. We argue that foresighting methodologies offer one approach for engaging communities in future-oriented thinking, debating, and shaping asset-based development. Foresight processes can empower and mobilize communities and aid in the creation of shared vision, strategy, and policymaking by seeking to better understand and even anticipate changing environments and complex public policy problems. Recent practice and scholarship have examined how foresight may be used to address innovation in specific sectors (Gaponenko, 2022). In Fall 2023, three key governance actors in the city of Corner Brook – Grenfell Campus of Memorial University, Port of Corner Brook, and City Hall – came together to explore alternative stories or potential future possibilities for the cruise sector in the community as for the city's main cruise-tourist corridor.

Global cruise tourism is a rapidly growing sector and Newfoundland is part of these global trends. Specifically, along Newfoundland's West coast, the Port of Corner Brook saw a 23% increase in total cruise ship passengers in 2024 compared to 2023 levels (Port of Corner Brook, 2024). Communities in Newfoundland have yet to take full advantage of cruise tourism, which requires a comprehensive and integrated long-term plan that considers the natural and built environment, or to fully examine the challenges associated with this global industry.

In Fall 2023, Grenfell Campus students worked with the City and Port to apply foresight techniques and a framework of drivers-scenario-action pathways to compare alternative future development possibilities for a tourist-friendly cruise corridor in 2035, as well as action pathways to achieve the desired futures for the corridor. A cruise corridor development plan was then prepared based on economic, historical, and inclusiveness perspectives and considering short-, medium-, and long-term potential pathways.

This exploration process took place as part of the GEOG 3350: Community and Regional Planning and Development class within the Grenfell City Engaged Learning Program, inspired by the global CityStudio model which brings together municipal officials and partner educational institutions to

*Figure 6.3* Scope of the project for tourist-friendly development of cruise corridor from the Corner Brook Cruise Terminal to Corner Brook City Hall. Credit: Jacob Moriarty, Aiden Callfas, and Dawn Quilty.

work together as part of coursework to co-create projects of community benefit. Previous topics taken on through this co-learning process have included animation of downtown space, sustainable waterfront development, outdoor recreation, and multi-modal transport planning. Similar projects are being undertaken in rural coastal communities (Figure 6.3).

## Public discourse and learning

Initial mapping can hint at the importance of public discourse and learning (both in the governance system and in broader social learning), which is our second infrastructure (see Chapter 3). An initial analysis might indicate that social learning requires bolstering since it currently revolves around a few topics and sites (Dunlop & Radaelli, 2020; Fischer, 2003). Governance learning might occur but on a fixed track. How can a community evaluate this situation more accurately?

Social learning is challenging to map. Yet it remains possible to identify the organizations, groups, initiatives, events, and places where learning is intense and where ideas are produced and reproduced (Harris, 1992). The same setting, professional group, or kind of gathering does not have the same learning aspects in each context. The *absence or scarcity* of public discourse or public media can make it harder for social learning to occur (Barnes et al., 2003). Even when it does, social learning can be limited to technical topics or might not actually question things (Luhmann, 2000).

Even where social learning (related to assets) entails some forms of comparative learning or dialectical learning, it does not guarantee that reflexive learning takes place. All forms of learning, however, can inspire rethinking. The root cause of problems, or a re-evaluation of overlooked strengths, in the social learning process might be found either in the realm of stories or in the infrastructures of learning. The problem might be related to the rigidity or the closed character of stories themselves and the connectivity of mutually reinforcing stories, or it might be in the world in which those stories are told, reproduced, discussed, and recreated (Bal, 2002). All these things can be examined and assessed. Locating problems in the realm of learning and public discourse can only come *after* analysis of the stories themselves (our first infrastructure). Possibly, what is missing is the persuasive performance of the stories, rather than features of the stories themselves.

If analysis indicates a lack of public discourse, social learning, and media, the matter could be that *spaces* are missing for this discourse to happen; opportunities, civil society organizations, or media outlets. It could be that widely shared stories about public discourse, social learning, or the role of organizations in society discourage meaningful reflection. There are, however, many other possible reasons for the lack of voicing, discussion, or questioning of assets, narratives, or community futures (Adamson & Donaldson, 2025). Media can be insolvent or captured by interest groups, public discourse might have been discouraged by a powerful elite in the past or felt unnecessary at a time when more unity and fewer problems were apparent (Schiffrin, 2021). Public discourse cannot be forced, just like meaningful participation. Previous chapters provided examples from Newfoundland and Labrador (NL) communities in which both problems and solutions were examined in the sphere of public discourse.

When a slow buttressing or reshuffling of the landscape of public discourse and social learning appears impossible, moves at a glacial pace, provokes excessive resistance, or demands significant investment, a focused intervention can offer a helpful alternative. It might be acceptable to organize said intervention for a period of time or with a focus on one topic that is accepted as worthy and in need of a thorough debate, as in the issue of food systems and food security in NL discussed by Emke below. Of course, paying close attention to the quality of public discourse and social learning is useful beyond the issue of reframing assets.

**Fit to eat: the Newfoundland farm and food show**

**A creative method to engage with local issues, strengthen public discourse, and encourage learning**

Ivan Emke

Many of us have devoted a significant amount of time to eating, as well as to talking, reading, and thinking about eating. However, we may have spent much less time considering where our food came from. Who grew it, raised it, hunted it, collected it… or made it in a lab? Questions around food production are particularly relevant for folks who live in NL, who are at the weak end of a transportation chain, all in an age of extreme weather, climate disruption, and corporate food system changes. There is a pressing need to increase food production and processing in the province. An important step in addressing this is to provide public settings to talk about food – how it is raised, produced, packaged, advertised, retailed, and sometimes driven across oceans, forests, and deserts to reach our plates.

But how do we start these conversations? As humans, we are storytellers. A barrage of statistics and expert clips don't have the impact or leave the lasting impression of a well-told tale. So the "Fit to Eat" project was essentially designed to tell some engaging stories about food and food producers, in order to contribute to the larger provincial food conversation.

Different communication tools can create different types of communities and can have different effects. However, there is a history of using radio to build knowledge about food and farming across Canada. Internationally, there is the continued success of Farm Radio International, which broadcasts programs throughout parts of the developing world. Radio has thus shown to be a useful tool for sharing stories about food. In NL, one of the most enduring radio programs on CBC is the Broadcast (originally called the Fisheries Broadcast), and in its early years, it included specific information about farming as well. On the other hand, communication tools such as Facebook and other social media can be prone to exaggeration and just plain nastiness – though there are a few notable exceptions of fine food production social media groups, managed by long-suffering and wise moderators (such as the NL-based "Backyard Farming and Homesteading NL"). I chose radio.

A goal of the "Fit to Eat" programs is to privilege the voices of everyday folks doing insightful things with food and farming. For the past 125 shows, we have featured conversations with over 140 farmers, processors, hunters, chefs, thinkers, innovators, homesteaders, researchers, cheese producers, earth-sheltered greenhouse builders, farmer's market operators, butchers, brewers, activists, food scientists, educators, policy-makers, gardeners, entrepreneurs, and eaters. Most are from NL, but the show sometimes goes further afield for interviews that will resonate with the NL context – including places like Iceland, Cape Breton, and even (trust me here) downtown Toronto. The program is currently broadcast on four community radio stations in NL (as well as stations in four other provinces). It is also available as a podcast on several platforms. Podcast listeners come from around the world. The updated programs are always available through the show's main website.

"Fit to Eat" takes a broad view of many aspects related to food in NL. There are some tough conversations, to be sure, but the overall tone of the program is optimistic and hopeful. We clearly find inspiration in the many women and men who work on increasing the food security of this province, and the show is dedicated to contributing to these conversations by listening to people who live and work – daily – with the challenges of feeding a province. It is a way to counteract the overwhelming sense of despair and futility that can come in the wake of awareness of huge problems, such as sustainably increasing food production.

One of the most important messages that I hope comes through in the shows is that everyone has a role to play in food security. Individual household percentages of local consumption vary widely from the provincial average, ranging from close to 0% to well over 90%. I want listeners to feel that small steps are the way forward. To put it simply, it may seem naïve to think we can save a province one carrot at a time, but at least that's a start – and it is far better than just giving up.

## Governance, reform, and experiment

As governance is both the system and the process that enables collectively binding decisions, it embodies the power to shape the community through those decisions (Van Assche et al., 2023). If it is difficult to articulate views of the future, discuss them, and learn about changing circumstances and policy responses within the governance system, the community has little power to choose a path and think in new ways about assets that enable change. Therefore, governance must be considered an important infrastructure for rethinking assets and building new futures on those assets.

If the stories people tell about their community, its future and past, good governance, and the good life are not impeding rethinking and transformation, and if public discourse is diverse and opens possibilities for rethinking what is of value, then the issue is not located in these infrastructures. If people take part in circles, events, or organizations where learning occurs rather than just instinctively reproducing fixed truths within a changing context, things may be working well in this domain. When narratives and public discourse are healthy, challenges might lie instead in governance. It can also be the case, however, that people feel dispirited after long years of inactivity in governance or repeated experiences of failed policies.

When we locate the root problem in governance, *a few typical situations* often occur. First, new stories in the community may find no response in governance or perhaps are purposefully ignored and blocked. Second, new perspectives may be noticed and provoke an internal response, yet a general capacity to organize (institutional capacity) is missing. A specific capacity to strategize might be missing, even where a rethinking of assets both in the community and in governance is taking place and institutional capacity is sufficient. Finally, there may be difficulty building new stories or new interpretations of existing ones. These different issues can combine, overlap, or sometimes attenuate each other. As we have seen, governance can entail many actors, institutions, self-assigned functions, and phases before something happens (Van Assche et al., 2021). Thus, the same issue can have many explanations.

If *new stories are deliberately blocked*, it can be because governance actors believe it goes against community values or notions of good governance, or would otherwise be problematic. It could mean that the community is not used to a response or that current governance actors see new ideas as threatening their interests. Political, administrative, or private actors that are influential in governance might be unable or unwilling to notice the changes in their environment, and thus not recognize the need for new stories. Administration might not lean heavily on participation or ask for input from citizens, or they might rely on past ways of doing things (Eakin et al., 2011; Scharpf, 1997).

If the issue is *institutional capacity*, it may be that new societal stories are recognized and understood, yet it is not clear which policy changes might be able to respond to these stories. Alternatively, ideas regarding appropriate responses may exist, but they cannot be implemented because other tools and governance infrastructures are missing. Specific *capacity for strategy* might be absent due to diverse causes, since various things make strategizing possible. There might be no tradition in articulating long-term perspectives in governance, or a dominant story that strategizing is not the job of existing governance structures, processes, and actors, and no capacity to question dominant stories like these. Alternatively, there might be hesitation in discussing or questioning others' perspectives due to uncertainty or fears of being perceived as disrespectful or offensive (Rahim, 2023).

Issues in *building new interpretations internally* can also be traced back to more than one issue. First, we emphasize that one cannot reduce governance to the translation of policy preferences in the community. Ideas can come from actors in the governance system itself, sometimes as a story combining aspects of several community stories, prompting governance actors to find a balance between desires and creating actionable and realistic options. Actors can also craft new stories, reinterpret assets, and open new futures, of which others can be persuaded. Leadership can be expected to play such a role.

This section represents a method for further investigating potential disconnects in governance once the problem of rethinking assets and futures is located there. The method can guide observation and analysis and will assist in the identification of additional methods. Surveys can be helpful if governance is felt to be unresponsive, but other useful methods include focus groups, which might mix actors both within the governance system and outside it, public conversations between governance insiders and civil society organizations, visiting critics, or engagement with internal outsiders. Beyond gathering information, it can be productive to find cracks in ossified discourse in and about governance, which can include the recognition of failure, omissions, blind spots, and internal contradictions. All of these can be useful for finding discourse that has hardened, locating difficulties for new stories to enter into that discourse, and identifying responses (van Zoest, 1986).

A potential result of this examination could be the recommendation of governance *reform*, which can take many forms. Introducing new paths of participation, making representation more comprehensive and fair, and pushing politics and administration to observe and analyze more seriously can improve responses, both on the part of individuals and through improved organizational structures and procedures. As noted earlier, reform might be discovered as desirable, yet not feasible at the moment.

*Experiment* might be in the cards, yet understanding available learning modes will be essential. One might want to consider who can observe and compare and who is permitted or has the capacity to draw lessons from the experiment. Conclusions of this self-analysis process might not question structures and procedures of governance, but rather performance, finding that more could be done. Perhaps an experiment that has been attempted in one place can be scaled up?

## Leadership

As we have discussed, sometimes the missing piece in rethinking assets and building futures is leadership. When (formal) leadership hasn't achieved much or discontent is widespread about current leaders, it can be easy to pinpoint the source of the problem. Often, however, it is only when other infrastructures have been ruled out that leadership comes into view as a concern. Prescribing appropriate behavior to leadership is extremely difficult. Thus, it can be hard to see the (potential) role of leadership and how that role could be enhanced. If leadership has naturalized its own role and the role of governance over a long time, imagining different models can be hard (Alvesson & Spicer, 2010).

In places where rethinking is needed but difficult, leadership can take on its most difficult, sensitive, and risky role: that of guiding reflection on the community's sense of self. Communities and their leaders may avoid this role as it goes against local pride and sensitivity, can unearth trauma, or expose fears. Holding up a mirror for the community at large, including leaders, to take a long, hard look at themselves is hard work that must be approached delicately. Governance may appear predefined or dominated by unquestioned voices, perhaps interacting in participatory settings or as technical deliberation. Understanding and dealing with trauma, supporting and nurturing new leadership, and facilitating conflict resolution are examples of required skills that leaders must have to be able to serve this role, but for which leaders are often ill-equipped. When formal leadership is the source of the knot, unraveling it is likely to require enlisting new leadership, people willing to step forward, and possibly move to formal positions. Even in distributed leadership, spaces for dissent often persist and must be protected and productively supported, along with spaces for mentoring new talent and subtle questioning of latent narratives, including narratives of governance.

In the following example, Shirley Montague illustrates how community leaders act in the role of builders (in this case crafting an event), while also creating a means for self-reflection and an asset for the community.

---

**Trails Tales Tunes (TTT) Festival**

Shirley Montague

Too often we find ourselves following the path of what we know rather than allowing ourselves to think outside the box. We tend to look outward rather than inward when it comes to measuring worth, be it tangible or intangible. Rather than seeing what we have, we tend to focus more on what we don't have.

Such was the quandary during the initial steps in building the TTT Festival (est. 2007) in Norris Point, NL, a town of approximately 700. Surrounded by Gros Morne National Park in Western Newfoundland, this town with stunning vistas from all angles was the inspiration and backdrop for a ten-day event during the month of May. The timing was unprecedented in the festival world of this province and was met with much skepticism at many levels.

It was in November 2006, when the seed was planted at a Town Council meeting, and with the Town's support, the work began. With a vision and mandate to showcase the town through a celebration of arts, culture and heritage, a group of community volunteers came together to launch a ten-day festival. Building on the appeal of the Park and all that comes with it, including a UNESCO World Heritage designation, the Festival started with a solid foundation. The event shone the spotlight on the town's amenities and residents, sparking a new community appreciation.

The ten days of programming included outdoor activities, and a variety of entertainment presented in multiple venues. It was a shift from a tent in a field or an event in a cold and vacuous stadium. It was an invitation for people to connect and celebrate. As it was an off-season event, businesses donated rooms to house festival performers and venues in which to stage the performances. Most events were admission by donation, making the Festival affordable to all. With a program that encompassed walks, musical matinees, activities for children, concerts, theater presentations, culinary offerings, workshops, and late-night dances, there was something for everyone.

The TTT Festival has become a signature event for the small town of Norris Point, elevating community pride and showcasing the town. Award-winning performers from around the world have shared the stage with locals and, while here, they become part of the community. A festival establishes a personality through its values and structure. Starting with community pride and a welcoming tone for all, the Festival is respected by performers and visitors alike and has become a much sought-after performance opportunity.

TTT has become the unofficial launch of spring tourism in the region, extending the season by several weeks. It is the impetus for accommodators and eating establishments to open their doors and for other businesses and operators to kick into action. Hence, the economic impact is easily calculated. The social impact is measured by uplifting smiles and a palpable vibe that says "welcome to this celebration of arts, culture, and heritage in our special little town."

## Toward strategy: diverse pathways in difficult conditions

Almost certainly, self-analysis will reveal several entangled issues across multiple infrastructures. Our basic model for strategizing (see Chapter 4 and later) moves from self-analysis to visioning and ultimately strategy. The conditions that make rethinking assets and futures necessary are often precisely why it is so difficult. In such situations, the path from self-analysis to strategy might not be a direct one, meaning that other things might be needed before articulating strategy becomes possible (Gruezmacher & Van Assche, 2022) (Figure 6.4).

Such *detours* for communities facing difficulties redefining assets and futures can entail:

1 *Institutional experiments*, in which the intention is to test something, draw the lessons, and decide whether the experiment can be generalized and become the normal form of organization. The experiment being tested can be a policy, the use of a kind of policy or institution, a form of coordination between institutions or actors, a new form of participation, a new emphasis on a form of expertise, a new arena for discussions of futures and assets, or a strategy.

2 A temporary form of governance, which we call *transitional governance.* Transitional governance can have the character of building institutional capacity in a limited manner without committing to a particular organization for the long term but rather improving conditions for long-term decision-making. Transitional governance can be more participatory than usual or can emphasize intensified learning, providing a context for self-analysis or perhaps providing a stabilizing solution for the short term.
3 *Institutional reform*, where a long-term structure for governance is envisioned. Possibly, this is also an exercise in building a platform from which more alternative futures become visible, but where the infrastructure is intended to be stable for the long term, allowing for the installation of new governance routines. Reform can include increasing participation, re-relating to external and internal outsiders, sharpening and multiplying observations of the environment, or whatever appears to be an obstacle in self-analysis, strategy, and ultimately action.
4 *Building assets*, perhaps where a new asset has been discovered, agreed upon, and seen to be helpful for strategy. Perhaps the community is considering a new piece of infrastructure, a new source of income for the municipality, or a utility company that will extend local autonomy – but no other assets or futures are visible yet. Self-analysis can continue in parallel, and obstacles in the consideration of assets can be revealed more easily when collaboration builds trust in one another and in the power of governance.
5 *Further analysis,* sometimes taking the shape of conflict resolution, may take place where neutral outsiders are brought in. Local leadership might work with mediators, facilitators, or consultants to work through collective issues, slowly and gradually open the interpretations of self, the environment, and the past, and from there recognize different possible assets and community futures. Elsewhere, it can be presented as extended visioning with a strong reflexive component.

Whether strategizing can start quickly after a knot has been untangled, or whether one of the detours discussed above is necessary, will likely become visible through the analysis of the infrastructures and obstacles toward collective strategy (Gruezmacher & Van Assche, 2022). Figure 6.4 illustrates a situation in which detours to strategy are common and diverse, while self-analysis can continue. Indeed, if rigidities in thinking and organizing are recognized, it would make sense to keep the conversation open and ongoing. Possible pathways between self-analysis, visioning, and strategizing are revealed by highlighting how certain aspects of self-analysis are more directly useful for strategy than others. It can also be helpful to show how some strategy problems can benefit from steps in self-analysis even when there is no perceived problem in the understanding of assets.

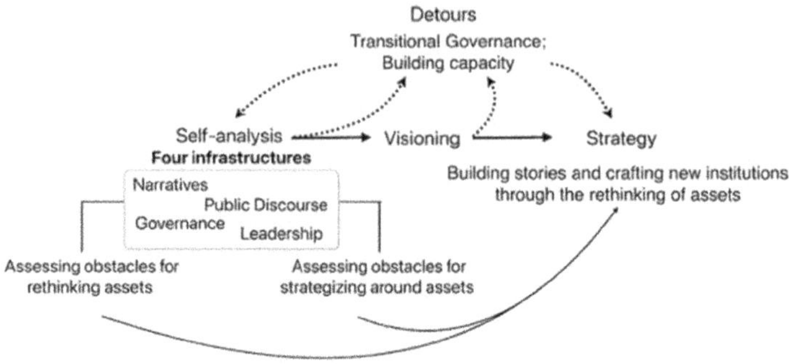


*Figure 6.4* Self-analysis and detours on the way to the crafting of community strategy.

## Concluding

Where the rethinking of assets is both necessary and difficult, the basic method of moving from self-analysis to visioning and strategizing still applies, yet self-analysis becomes more important and can be organized around a careful study of four infrastructures for reconsidering assets: narratives, public discourse and learning, governance, and leadership. These infrastructures shape each other, as leadership requires governance and vice versa, and as stories and modes of organizing are intertwined. Understanding and possibly mending these infrastructures can enhance flexibility in the collective understanding of self and environment, both past and present. This in turn can prompt new stories about the future and the assets that such futures demand.

A second feature discerned in communities facing difficulties in rethinking assets and futures is that the paths from self-analysis to strategy become increasingly diverse. Self-analysis can uncover issues in the framing of assets and possibilities for change, but it can also reveal problems in strategizing. The more work is invested in the rethinking of assets, the more important it becomes to achieve several goals through both laborious self-analysis and enabling strategy. In these situations, it is important to avoid artificially separating extensive self-analysis from considerations of strategy (Van Assche et al., 2019).

Each infrastructure thus must be well understood, and in mapping narratives, assets, learning, and governance paths, it is essential to assess the extent to which the four infrastructures contribute to rigidities in the understanding of assets and futures. This is the overarching method suggested here. This self-assessment will identify points of intervention, modification, and ultimately transformation. Governance processes and structures, both existing and newly created, both use narratives as inputs and produce narratives as outputs. They have the power to produce new stories and new futures, thereby making them real.

Where positions seem immovable, even when existing realities are falling apart, time might do its work. Immigration, ecological and economic shocks, or shifts in multi-level governance can open thinking, but so can storytelling, art, travel, and visitors or newcomers. Admiration by visitors, finding a place in different networks of learning and entrepreneurship, rediscovery of different pasts by local performers and writers, disillusion when visiting other places, confusion when confronted with art, or being charmed by outsiders hiding in plain sight can all introduce shifts in perspective which can gradually transform the community's self-understanding (see annex). However, by this point we are no longer in the realm of method and strategy, but in the world of social and cultural change which, in the long run, leaves no story unaltered.

## References

Adamson, G., & Donaldson, L. (2025). Schizophrenia: Proportionality and erasure in Canadian news media. Journalism, 14648849251319196. https://doi.org/10.1177/14648849251319196

Alevizou, G., Alexiou, K., & Zamenopoulos, T. (2016). *Making sense of assets: Community asset mapping and related approaches for cultivating capacities* (pp. 43–43). The Open University and AHRC.

Alvesson, M., & Spicer, A. (2010). *Metaphors we lead by.* Routledge.

Bal, M. (2002). *Travelling concepts in the humanities: A rough guide*. University of Toronto Press.

Barnes, M., Newman, J., Knops, A., & Sullivan, H. (2003). Constituting 'the public' in public participation. *Public Administration, 81*(2), 379–399.

Brown, C., & Augusta-Scott, T. (2006). *Narrative therapy: Making meaning, making lives*. SAGE Publications.

Buchinger, E. (2007). Applying Luhmann to conceptualize public governance of autopoietic organizations. *Cybernetics & Human Knowing, 14*(2–3), 173–187.

Czarniawska, B., & Gagliardi, P. (2003). *Narratives we organize by* (Vol. 11). John Benjamins Publishing.

Denscombe, M. (2024). Decolonial research methodology: An assessment of the challenge to established practice. *International Journal of Social Research Methodology, 28*(2), 231–240. https://doi.org/10.1080/13645579.2024.2357558

Dunlop, C., & Radaelli, C. (2020). The lessons of policy learning: Types, triggers, hindrances and pathologies. In G. Capano & M. Howlett (Eds.), *A modern guide to public policy* (pp. 222–241). Edward Elgar Publishing.

Eakin, H., Eriksen, S., Eikeland, P.-O., & Øyen, C. (2011). Public sector reform and governance for adaptation: Implications of new public management for adaptive capacity in Mexico and Norway. *Environmental Management, 47*(3), 338–351.

Fischer, F. (2003). *Reframing public policy: Discursive politics and deliberative practices*. Oxford University Press.

Gaponenko, N. (2022). In search of sectoral foresight methodology: Bridging foresight and sectoral system of innovation and production. *Futures, 135*, 102859. https://doi.org/10.1016/j.futures.2021.102859

Gordon, J. (Ed.). (2012). *Community radio in the twenty-first century*. Peter Lang.

Gruezmacher, M., & Van Assche, K. (2022). *Crafting strategies for sustainable local development*. InPlanning.

Gunder, M. (2015). The role of fantasy in public policy formation. In R. Beunen, K. Van Assche, & M. Duineveld (Eds.), *Evolutionary governance theory: Theory and applications* (pp. 143–154). Springer International Publishing.

Hahn, T., Schultz, L., Folke, C., & Olsson, P. (2008). Social networks as sources of resilience in social-ecological systems. In Jon Norberg and Graeme S. Cumming (Eds.), *Complexity theory for a sustainable future* (pp. 119–148). Columbia University Press.

Harris, E. M. (1992). *Dreaming reality: Small media in community development as critical educational practice. A case study of community narrowcasting in the Town of Buchans, Newfoundland, Canada* [University of Toronto]. https://eric.ed.gov/?id=ED373194.

Howley, K. (2005). *Community media: People, places, and communication technologies*. Cambridge University Press.

Jeppesen, S. (2016). *Understanding alternative media power: Mapping content & practice to theory, ideology, and political action. Democratic Communiqué*, *27*, 54–77.

Kapitan, L., Litell, M., & Torres, A. (2011). Creative art therapy in a community's participatory research and social transformation. *Art Therapy*, *28*(2), 64–73.

Lowe, S. S. (2000). Creating community: Art for community development. *Journal of Contemporary Ethnography*, *29*(3), 357–386.

Luhmann, N. (2000). *The reality of the mass media*. Stanford University Press.

Mintzberg, H. (2017). *Crafting strategy*. Routledge.

Mohatt, N. V., Thompson, A. B., Thai, N. D., & Tebes, J. K. (2014). Historical trauma as public narrative: A conceptual review of how history impacts present-day health. *Social Science & Medicine*, *106*, 128–136.

*Port of Corner Brook Experiences Another Record-Breaking Cruise Season – Corner Brook Port*. Retrieved February 18, 2025, from https://www.cornerbrookport.com/port-of-corner-brook-experiences-another-record-breaking-cruise-season/

Rahim, M. A. (2023). *Managing conflict in organizations* (5th ed.). Routledge. https://doi.org/10.4324/9781003285861

Scharpf, F. W. (1997). Introduction: The problem-solving capacity of multi-level governance. *Journal of European Public Policy*, *4*(4), 520–538.

Schiffrin, A. (Ed.). (2021). *Media capture: How money, digital platforms, and governments control the news*. Columbia University Press.

Steiner, L. (2012). Community journalism must tackle tough local issues. In B. Reader & J. hatcher (Eds.), *Foundations of community journalism* (pp. 21–23). SAGE.

Van Assche, K., Beunen, R., Gruezmacher, M., Duineveld, M., Deacon, L., Summers, R., Hallstrom, L., & Jones, K. (2019). Research methods as bridging devices: Path and context mapping in governance. *Journal of Organizational Change Management*, *34*(4), 729–746.

Van Assche, K., Gruezmacher, M., Vodden, K., Gibson, R., & Deacon, L. (2021). Reinvention paths and reinvention paradox: Strategic change in Western Newfoundland communities. *Futures*, *128*, 102713–102713. https://doi.org/10.1016/j.futures.2021.102713

Van Assche, K., Gruezmacher, M., Marais, L., & Perez-Sindin, X. (2023). *Resource Communities: Past Legacies and Future Pathways*. Taylor & Francis. https://books.google.ca/books?id=4EHUEAAAQBAJ

van der Watt, P. (2018). Community development in wounded communities: Seductive schemes or un-veiling and healing? *Community Development Journal*, *53*(4), 714–731.

van Zoest, A. J. A. (1986). Zin zien: over interpreteren. Muiderberg: D. Coutinho.

Wardle, B. (2016). You complete me: The Lacanian subject and three forms of ideological fantasy. *Journal of Political Ideologies*, *21*(3), 302–319.

Weisel-Barth, J. (2019). Stories that open and stories that close: Theoretical and clinical narratives in psychoanalysis. *Psychoanalytic Inquiry*, *39*(7), 485–493.

Žižek, S. (2017). The seven veils of fantasy. In Dany Nobus (Ed.), *Key concepts of Lacanian psychoanalysis,* (pp. 190–218). Routledge.

# 7 Conclusion

## Creating spaces for rethinking development paths

**Abstract**

*The hundreds of outports of Newfoundland and Labrador, and its much smaller number of inland communities, have endured dramatic changes in recent decades, but also over centuries of colonialism that have given the province a unique experience. At the same time, this history shares similarities with other places which can serve as a warning for troubles coming, an inspiration for other rural or island regions, and a reminder of causal links that might have been forgotten. The Province of Newfoundland and Labrador was incorporated late into Canada, prior to which it remained in a colonial status within the British Empire for almost a century. Of course, colonialism endures today with the continuing legacy of displacement of Indigenous peoples, traditions, and cultures. This colonial legacy has also affected settler Newfoundlanders and Labradorians, albeit in a different way, wherein the British government and merchant families exercised power over communities but did not invest adequately in them, instead concentrating administrative and political development in the colonial capital of St. John's. This core-periphery dynamic continues to impact the contemporary institutional landscape of the province.*

In this book, we offer a methodology for the self-assessment and discovery of community-based assets, in service of building community strategies that enact reflections from this re-interpretation of community and its environment. The methodology, which aims to dislodge difficulties and unravel knots preventing a rethinking of strengths, is inspired by this history of challenges and solutions in a Canadian province that has in many ways and over time

DOI: 10.4324/9781003620297-7
This chapter has been made available under a CC-BY-NC-ND 4.0 license.

had to re-evaluate its own assets and relationship to the environment, aiming to glean lessons from place-based solutions that communities have attempted and struggles to overcome the hurdles encountered along the way.

Newfoundland and Labrador communities have been striving over centuries to overcome trauma, shocks, and the limitations stemming from this colonial history of neglect and marginalization. Numerous are the stories of determination in communities fighting to overcome obstacles and find new solutions, providing a parallel history of community-based experiments, successes, and failures with many examples of learning, modifying, and changing course along the way. Many of these local experiments involve governance and the institutional landscape in some way. All of them have something to offer for other communities grappling with major shifts in their underlying assets, narratives, or governance structures, which this book aims to share while considering lessons learned and sharing vignettes that tell the rich stories of these local experiments and experiences.

If communities feel very comfortable in their reality, or if that reality feels threatened, they may cling to it. Unwillingness to question the viability of current assets can stem from a drive to preserve activities that once sustained a community but may not be as central in the future. Spaces for fruitful self-reflection and critical discussion are needed. In a democratic setting, communities decide for themselves what they consider is a strength or asset upon which futures can be built, which pasts and futures they believe in, and how they want to move forward. In the same settings, however, many feel the limitations of their existing forms of organization, values, and narratives, a feeling that existing economic assets are no longer sufficient and that other local assets, however they are valued, may not be of sufficient economic significance to ensure community survival. Of course, economic values are not the only factors that matter for community sustainability and resilience, but they are an important consideration nonetheless, particularly in times of economic decline.

## A self-analysis to find new flexibility in thinking and organizing

Hence, in the previous chapters, we distinguished four infrastructures for rethinking assets and building community strategy around them, infrastructures which can be present, absent, functional, or dysfunctional to different degrees.

1. *Narratives*, with a special role for identity narratives framing assets.
2. *Public discourse and forms of learning*, with a special interest in dialectical learning and diversity of perspectives.
3. *Governance*, with a special interest in governance reforms and experiments for building institutional capacity.

4 *Leadership,* with a focus on distributed leadership and a specially articulated set of leadership roles where sustainability threats occur.

This book argues that mapping, understanding, and improving these infrastructures *is the key method* for finding new flexibility in thinking and acting as a community. The main obstacles and enablers can be found in different infrastructures in each case, while several infrastructures can combine to create either more or less favorable conditions for rethinking community assets. This approach can also be thought of as an umbrella method that allows for the identification and implementation of many other methods (Figure 7.1).

For understanding all four infrastructures it is first useful to have a basic understanding of the history of the community and its governance system. Tracing its evolution is valuable for understanding the present, even in cases when governance was largely informal until recently. Second, an initial mapping of assets and key narratives is helpful. Here we speak of stories about the past, present, and future, relations with the environment (Alevizou et al., 2016; Mintzberg, 2017). Mapping governance helps us understand where institutional reform or building might be needed and where learning and non-learning occur.

*Figure 7.1* Back cover of a tourist brochure. Newfoundland. Tourist and Publicity Commission (1937?) Newfoundland: Britain's oldest colony where sport is at its finest [booklet/text] Centre for Newfoundland Studies – Digitized Books (file name 140682.jp2).

Meanwhile, mapping narratives and assets is helpful for all four infrastructures since a community is ultimately a way people organize themselves and understand themselves through stories. Organizing and thinking affect each other in myriad ways, hence the importance of understanding both and their intersections in governance (Lowery et al., 2019; Van Assche & Gruezmacher, 2022).

Each infrastructure thus must be well understood and the overarching method entails conducting a self-analysis of the contribution of the four infrastructures to rigidities in the understanding of assets and futures. This analysis identifies points of intervention. Each infrastructure has a different susceptibility to intervention, fixing, mending, and ultimately transformation. As we have discussed, governance uses narratives as input and produces them as output, creating the ability to produce new stories and new futures – and to make them real.

## Good governance as creating and maintaining flexibility through diversity

Self-analysis can point at a variety of issues. As it is carried out by a community, it is not up to the authors to give recommendations on the direction of strategy. What we can do is distill from our previous observations on the four infrastructures a set of general principles regarding good governance, about features of governance systems that support a more flexible yet fair approach to assets and strategy.

*First of all*, governance should consider a wide variety of truly different perspectives on the past, future, environment, and self, as well as foster and preserve the tools to assess them, typically involving both community members and outsiders.

*Second*, governance should maintain and further develop the capacity to learn in all its forms and mobilize this learning for the creation of new (or in some cases renewed) narratives.

*Third*, governance should maintain and enhance the capacity to strategize, which entails the construction of new or consciously renewed narratives, and a deep toolbox of institutions and instruments that can be matched to the guiding narrative.

*Fourth*, both people and their stories need to be connected in as many ways as possible to governance, through a combination of participation and representation that serves to maximize the input of stories. Possibly, a phase of more intense participation in special events and venues is useful when public discussion about stories and assets needs to be reopened or expanded.

*Fifth*, improving the diversity of perspectives and reflexivity in governance requires action. Redefining insiders and outsiders, where possible, introducing outsiders or re-opening relations with distrusted external actors (on the community's terms) can contribute to increasing this diversity and reflexivity. So can bringing in more diverse insider perspectives, perhaps individuals and groups not yet invited to share their perspectives. From there, dialectical learning and truly new perspectives can emerge.

Communities deserve the opportunity to reimagine the value of the past while identifying new pathways forward based on the foundation of assets cherished, revealed, or constructed. Each redefinition of assets and new perspective on the future will retain traditions, values, narratives, and assets, but in a new combination. Thinking in terms of binaries and all-or-nothing options hinders change and reduces diversity in perspectives. Genuine assessment of assets and imagined futures becomes almost impossible when the strategy is reduced to a choice between past and future.

What method(s) work for a given community to reconsider their assets and build strategy will generally require somewhat of a trial-and-error approach. We draw inspiration from the experiences of Newfoundland and Labrador communities to that end – their challenges, their diverse and courageous attempts, and their experiments to overcome these challenges, particularly tough where a central asset has been degraded or devalued. We have shared examples of strong economic and political leadership, creative institutional bricolage and building of institutional capacity, efforts to reveal beauty, rich cultures, and traditions, to find new resources in old landscapes and new uses for old resources, and more. Many community development experiments and trajectories shared in this book illustrate the important role of the arts in reaching people's hearts and allowing them to see a side of themselves of which they were not aware. The examples featured here also show how the past, from a distant geological past to more recent history, can serve as inspiration for rethinking assets and futures, yet it can also constrain creativity if tied into identities that are not receptive to change or evolution.

The myriad small communities represent a mode of adaptation themselves, as their scale and location have shifted along with policy and understandings of the number of people needed to survive together. Factors such as adjacency to fisheries resources, the resources and new levels of investment required as service expectations change, new technologies, investments, connectivity, changing markets, and methods of seafood processing have made living in some small remote communities less necessary and less attractive for many. As seen in Chapter 1, top-down policy initiatives and modernist narratives employed by the Smallwood government pushed for the centralization of the population and resettlement of more remote outport communities. New ideas and ideals were introduced, through modern media or transportation connections for example, expectations and understandings of quality of life changed, and it was possible to piece new livelihoods together as lifestyles became, for some, less rooted in place. New connections were created for the exchange of goods and travel to larger centers, services become more accessible, virtual connections more commonplace, and technology now allows for exploitation of resources further away. Change continues, as does the need for reinvention, as it does.

In Newfoundland and Labrador and elsewhere, these changes have reduced the need to stay in a place for the resources or livelihoods it provides, while on the other hand increasing possibilities to stay in a place if one finds it attractive. Individuals can move more freely, and communities have more options to do other things. At the same time, for new residents to move in elsewhere requires a quality of life and supporting infrastructure that enables both work and daily living (high-speed internet for mobile professionals, for example). These trends have been observed around the world and long studied by sociologists and geographers, which can potentially inform communities in exploring new possibilities, along with their own reflections and experiences.

## Networks and multi-level governance

What is further in line with observations elsewhere, and can strengthen the case for community self-assessment, is that local action must adapt to macro-level forces. Expectations are higher, networks are larger, and supply chains are more complex today than in the past, implying that maintaining a traditional activity requires continuous innovation. These realities make the supra-local level more important, whether in the form of (sub-)regional governments or intensive collaboration between municipalities and other local governance actors. Regional governance can be promoted for specific services or policy domains. A collection of small, isolated towns is likely to face tougher challenges in the future than a network of small towns well connected through strong collaboration or regional governance.

Well-connected rural places have the potential to be more attractive to new residents than larger centers with a more homogeneous character. For example, the recognition of functional regions in policy frameworks can make all the difference, not only for the affordability of service delivery and infrastructure in areas like healthcare, elderly care, education, water, and waste management but also for an ecosystem of diverse professional services to thrive (Freshwater at al., 2013). Key services for young professionals like daycare, dog parks, or recreational amenities can heavily influence individual decisions of whether to move to a community, which rural communities may need to pool resources to provide. The result can be a web of small settlements that gradually come closer together in the sense of functional integration but still maintain distinct local identities that contribute to the attractiveness of the place for new and existing residents.

Many communities in Newfoundland and Labrador are enhancing connectivity, but there must also be effective leadership and institutional capacity to take full advantage of opportunities. This means that local government must expand its mandate beyond providing basic services. Articulating futures for the local community, finding policy tools and resources to turn a shared vision into a strategy, developing active networks of collaboration at local and higher levels, and understanding and communicating the community's quality of life and amenities to potential new residents and businesses are necessary to turn a set of small communities dependent on one or few assets into a networked region where assets can be discovered, attracted, and assessed in new ways.

Therefore, in places where no meaningful local governance capacity exists, it must be developed, and where local governments are only able to do the bare minimum, there must be efforts to strengthen local resources and capacity in a manner that centers local priorities but also finds support in senior government policy frameworks.

On the other hand, higher-level government cannot simply dictate orders to the local level, and communities should be wary of "devolution" efforts that actually download responsibilities onto communities that they do not have capacity to deliver. To address the wicked problem of transitioning away from a resource that is no longer viable and on which under-resourced communities may depend, a multi-level approach to governance is needed. More precisely, the governance system must make space for the articulation of local and regional futures. Community-level voices must inform the articulation of regional futures, while simultaneously developing institutional capacity at the regional level and a regional identity that draws from local identities and narratives. Otherwise, the region is likely to be no more than an arena for competition between communities over scarce resources, or a place to dream up wish lists without the capacity to pool resources for collective action to realize those dreams.

## Inclusivity, oppression, and reinvention

In discussions on assets and visions for the future, there must be voices that represent all levels of governance at the table, not only to help maintain checks and balances and provide an outside view but also in recognition of public goods and assets at each scale and to draw from the available assets and resources of each level. This multi-level approach both sharpens the ability to test the feasibility of visions and strategies at each level and increases the diversity of perspectives around the table. No community can be completely autonomous, and no small difference is worth isolating oneself from neighbors and potential partners and in turn diminishing local capacity for adaptation and reinvention.

In Newfoundland and Labrador, and elsewhere, it is clear not only that certain actors and their voices have been excluded for many generations but also entire traditions of governance. Indigenous people were not recognized when the colony of Newfoundland became a Canadian province, their forms of governance were neither observed nor supported. Therefore, learning from those governance traditions did not take place, and many adaptive practices were not understood. Meaningful forms of collaboration between settler and Indigenous communities were hard to develop in this context, given this fundamental misrecognition, and traumatic history of forced assimilation and relocation.

Indigenous actors are increasingly being recognized as relevant and necessary governance actors, and an important part of local, regional, and provincial governance systems throughout Newfoundland and Labrador. The province's early colonial history has undoubtedly impacted the way that Indigenous peoples and their systems of governance were and are exercised. But this does not mean that these forms of governance do not, or have not, evolved over time. Despite pervasive colonial acts that have attempted to undermine and render them obsolete, Indigenous systems of governance are becoming revitalized and reclaimed by Indigenous leaders and community members. Albeit, these forms of Indigenous governance, while deeply rooted in the past, also represent a relatively recent contribution to how governance is exercised in the province. This reality has the potential to position communities to envision new futures, leveraging diverse approaches to governance.

## A word of caution, a word of hope

Strong local identities can strengthen communities but also sometimes make it harder for open discourse and create divisions between local narratives on past, future, the environment, and assets. Consider a small Catholic community whose neighboring village 10 kilometers away has many shared experiences, but since its residents are Protestant, there is minimal shared connection and lack of trust. Identity-based divisions can hamper collaboration with neighboring communities and other external actors and reinforce things remaining exactly as they are. The same two communities may be perfect candidates for sharing public infrastructure such as water treatment or recreational services, but local leaders find it unthinkable to cross the religious and cultural divide to collaborate.

What may seem like a small difference can be felt as essential, and small changes to cemented identities and ways of doing things can be perceived as a major disruption or threat. Rigid identity narratives can keep a cluster of other narratives in place, which together create conditions where rethinking assets is very difficult.

When histories have been difficult, traumatizing, and characterized by shocks to community survival, strong emotions are likely to arise whenever the past or the future comes into focus. Some aspects of the past may even be inaccessible or off-limits for discussion. A traumatic past of exploitative resource extraction, colonial oppression, or shocks to local economies often combine to produce assessments of the future that can be either overly fearful or positive.

Institutional capacity should not exist in and of itself but in support of a governance system where reflection about identity, past, and possible futures, becomes possible. A tailored capacity enables the community to assess potential futures in a way that neither clings to the past nor bows to the priorities of the long-standing elite but truly reimagines past and present assets in a way that could lead to transformative futures. In such a governance system, reflexivity, second-order observation, and true diversity of perspectives must be supported, as they underpin all forms of learning. Together they can slowly thaw out frozen interpretations of assets and futures or unravel the knot of stories that keep old assets tied up and unable to be reused. If outsiders have a hard time being accepted as insiders, and insiders are turned into outsiders for increasingly trivial reasons, diversifying observation in and of governance becomes an urgent matter. Self-analysis can assist in uncovering stories and breaking the impasse.

Leadership might be tempted to see these efforts as peripheral to the future of the community, particularly when the hierarchy of needs requires a focus on survival and providing basic services. Discussions of long-term futures may seem like a luxury the community cannot afford. Leadership may feel threatened by efforts to rethink narratives that could destabilize their regime. However, where non-learning dominates governance or the nature of dominant narratives narrows observation and self-observation to the extent that only a few problems and solutions are acceptable, self-analysis is not a luxury – it can be a matter of life and death.

What we propose is burdensome and risky. Communities must decide for themselves how much they are willing to invest in a collective reflection that is potentially destabilizing. Some may not be ready for candid reflection, some might be unable to devote the time and effort required. In considering whether such a self-analysis is right for them, communities can first ask themselves if their governance routines are able to avert hardship, particularly in the face of the realities of constant and often increasingly rapid change. If not, this may be a good sign that a community could benefit from this difficult but rewarding journey. Even after a community arrives at a collective strategy, in and through a reinforced governance system, it can sink back into an inflexible paradigm. We hope that the methods of self-analysis, and stories of community creativity and experimentation in Newfoundland and Labrador shared in this book can help to keep the potential for new ways of thinking about communities alive even in these situations, offering new ways of thinking about community assets and futures and practical guidelines for communities to consider how to embark on a journey of self-reflection and futuring that makes sense for their unique assets, histories, and values.

## References

Alevizou, G., Alexiou, K., & Zamenopoulos, T. (2016). Making sense of assets: Community asset mapping and related approaches for cultivating capacities (pp. 43–43). The Open University and AHRC.

Freshwater, D., Simms, A., & Ward, J. (2013) *Functional regions as a structure for enhancing economic development in Atlantic Canada: Background Report.* Project Report. Memorial University of Newfoundland, St. John's, Newfoundland.

Lowery, B., Vodden, K., & Campus, G. (2019). *Analysis of asset mapping initiatives in rural NL.* ACOA.

Mintzberg, H. (2017). *Crafting strategy.* Routledge.

Van Assche, K., & Gruezmacher, M. (2022). Asset mapping 2.0; contextual, iterative, and virtual mapping for community development. *Community Development, 55*(2), 195–210. https://doi.org/10.1080/15575330.2022.2131861

# Annex
# A creative catalog of methods

The methods presented here are generally meant to serve as part of self-analysis and strategy. They can assist the broader method developed in Chapter 6, for diagnosing difficulties in the reframing of assets, and the building of community strategies. Self-analysis, in communities finding themselves tied to assets that do not support the community to the same extent they used to in the past, can be structured around the investigation of four infrastructures usually enabling the reframing of assets. Some methods presented here are more about finding facts, which can assist in decision-making, others about finding stories, which might shed new light on the community, and some are more experimental tools of governance, still aiming at enhancing reflexivity, diversity in perspectives, and shifting interpretations. Most of these methods have been illustrated through the vignettes in this book, providing practical examples. Finally, we note that nothing stands in the way for the reader to devise their own methods, to add to the list, and to combine things into a method of self-analysis, visioning, and strategy, which is entirely idiosyncratic, that is, tailored to the needs, possibilities, and identity of the community.

## Art and interpretation of the familiar

### *Community theater*

Community theater can be a hobby of a few local residents and it can be a powerful tool to entertain, collaborate, to explore different perspectives, for the actors and the spectators. Plays can be short or long, can be stand-alone or part of bigger cultural events, taken from the standard repertoire, or, perhaps more interestingly, written or rewritten by locals to comment on current events, episodes in the life of the community, and challenges it faces. Especially identity narratives can be explored, sometimes directly by talking about the community, sometimes critically reflecting from different angles, through the eyes of different characters, and maybe more easily by talking about similar issues in other times and places. If things get too didactical, people tend to lose interest, so rather than teaching and criticizing, a meaningful role for

*Figure A.1* Community theater. "The 39 Steps" directed by Todd Hennessey for "Stage West Theatre Festival" 2012. From left to right, Greg House, Mike Payne, Steve Perchard, and Rebecca Pike. Credit: Jacob Hennessey.

community theater (as opposed to just theater) in our view associated more easily with storytelling, remembering, depicting, and exploring (Figure A.1).

### *Festivals and other larger events*

Festivals take time to organize and require resources. They can also bring a community together in an intense manner, as both planning and improvisation are needed, strategy and tactics, mutual support, and the atmosphere during the festival can shed an entirely new light on the community. A festival can highlight both local and other stories, products, and qualities, and can be conceived with both locals and outsiders as audience. Thus, a festival can contribute to economic development (directly) and it can reshape the image of the place, especially if helped by media coverage and tourism networks, and create longer-lasting brand value. Even where very few activities question anything in local tradition, where the festival hopes to preserve tradition and social bonds, it likely offers occasions to reflect on those traditions and the nature of those bonds.

### *Public art*

Public art can also be a most powerful way to make people observe, think, and feel differently about themselves, their community, and their environment. If artwork is mostly the expression of feelings about something entirely unrelated to the community, it might inspire pride of place, but will not have much of an effect in terms of shifting stories and feelings about the community itself. If dialogue between the artist, commissioning organization (where relevant), and community can precede the purchase, the process can increase the chances that people pause and think. The work might become a marker of local identity and incite local pride. Public art can also be temporary and/or a collective and inclusive effort.

***Images: photo and video***

Photos can be exhibited, videos can be shown in public, they can be presented as art, as documentation, and they can feature in other ways in methods of research and methods of governance. Photos can be used in visioning sessions, in self-analysis, in discussing a particular controversial project, old photos can spark discussion about the identity of a place, about local qualities and assets. Strategies will most likely include images, making it easier to imagine the kind of community that could materialize. Video can convincingly convey a slice of life in the community, past or present, and depict its environment and the use of resources more comprehensively. Techniques such as photovoice can elicit video from research participants but also participants in governance in the case of visioning and strategy processes. Images can open up discussion by revealing a reality that contradicts common narratives.

***Writers or musicians in residence***

Writers, musicians, and poets come to appreciate a place in their ways. The presence of an artist can give cachet to a place, or reinforce an existing local taste for culture; it can be interesting for visitors and locals, who might enjoy not only the results of the stay but also the presence of the artist, casual conversations, or some more serious lectures or performances. There could be an encouragement to engage with the place, its stories, and values. Literary festivals can connect communities with visitors, can combine with exhibitions or other events, and can change the image of a place for outsiders and for itself. The alienation from everyday meanings that poetry can bring, might yield new meanings, and new connections.

## Investigation for reinterpretation

### History and memory

History and memory are written and rewritten all the time, so it might appear strange to see them as methods. Knowing one's history is a double-edged sword: one can become trapped in fixed stories that ought to be questioned and opened for reinterpretation, but simply forgetting is not advisable either as the past can orient us in the present and help devise futures. If a community is interested in revisiting its future, it makes sense to revisit the past. Discussing the reasons for consequential choices and revealing those consequences can be productive in revealing what had been naturalized over time. Interrogating history in as many ways as possible, using archives and fully investigating social memory, might be useful in finding new ways to understand the centrality of an asset, and in some cases moving beyond it.

### *Mapping and moving*

Collecting maps, old and new, general and thematic, making maps, making maps collectively, and using maps collectively are excellent ways to see a community anew. Participatory mapping of heritage, of naturally, esthetically, or socially important sites, can spark new conversations in a group, and bringing in an expert can further deepen such conversations. Touring or mapping can focus on marginal areas, on forgotten heritage, or on the histories of people not living there anymore. This observation, conversation, mapping, and their interplay can lead to new stories about place and community.

### *Asset* mapping

Asset mapping, making an inventory of what is of value for the community, is helpful mostly as an introduction to other forms of self-analysis, and much of this book deals with situations where an initial asset mapping must be treated with caution, where assets require reframing, but where this proves difficult. The overarching method offered in this book is one where self-analysis tries to locate the knots in stories about assets and futures, aiming to diagnose the obstacles for a reinterpretation of assets. We identified four infrastructures for the reframing of assets, where such issues could be found: narratives, leadership, governance, and the realm of public discourse and social learning.

### *Citizen science*

What is called citizen science is often a mobilization of interested citizens in larger-scale research sponsored or encouraged by academic or government institutions. Citizen science can be participatory, taking the form of any kind of investigation that is helpful for self-analysis, visioning, and strategy. The kind of data and materials collected, observed, counted, and remembered will depend on the interests of the people involved as well as the expertise available. Citizen science makes sense as an input not only for local governance but also to bring people into the orbit of governance and public discourse, and it provides a way to refocus stories about the community, through detailed observation, which can in turn alter the experience of place. For the community, the sum of all details can result in a new picture or topic of interest for strategizing.

### *Narrative mapping and path mapping*

Mapping of key narratives in governance and communities and mapping of governance paths are umbrella methods, useful for self-analysis and strategy (see Chapter 6).

## Confronting and rethinking

### Comedy

Comedy cannot be forced, and comedy cannot be a bureaucratic exercise. Communities, however, do have their comedic traditions and sensibilities. And often these involve holding a mirror to power, to the community itself. Nowadays, stand-up comedians and others with less standing reflect on culture and politics and show the vulnerabilities of both themselves and the societies they live in. Before dismissing this method entirely, maybe we should remind ourselves that, before the bureaucratization of local governance (and even after), gentle and not-so-gentle mockery was often part and parcel of local culture. Mockery helped to bridge the divide between factions and between leaders and the rest – especially when they were not that different from each other.

### *Roleplay*

Like community theater, roleplay can be an interesting way to explore different perspectives, yet in a smaller format, and usually less scripted. Like community theater, it tends not to work if people feel lectured, shamed, or coerced. Roleplay must be incorporated into other activities, and these can be linked to either self-analysis or strategizing. With a skilled mediator, it can be part of conflict management. In a more light-hearted version, it can be a skit or part of a larger event. Roleplay can also (in self-analysis) be used in a more structured way, to explore collectively a problem or opportunity, as a mode of social learning.

### *Least beloved creature competition*

Interesting for locals and visitors, alike, maybe starting from children, but involving others later, this could be a good way to revisit the natural environment, and the relation to and use of the landscape. The intention is to agree on elements of the surroundings that are less loved and reflect on why and what is at the root of the aversion. The competition could be open to stories, drawings, songs, photos, or combinations of these, and could be coupled to an exhibition and discussion. A more innocent version could revolve around creatures most hidden and unknown.

## Organizing and opening perspectives

### Participatory budgeting

Here budgets are not decided by top management (such as council and administration) but at least in part at lower levels. In an organization, this can mean that middle managers propose their budget to higher management, or that employees are directly involved. In communities, it can mean that administration is more broadly involved, and most importantly it might say that citizens are directly making budget decisions. This can be empowering, as locals can feel aligned and may be identified with their local government. It can enhance transparency, as old priorities might become visible and understandable, and as the incompatibility of desires might reveal itself. Participation can take place through meetings, public debate, special committees, and surveys. Local media can play a role, by presenting and discussing different proposals, coming from council, administration, or citizens. When not framed by a vision or visioning process, participatory budgeting can play into the emotions of the moment and shorten time-horizons, consolidating rather than interrogating dominant narratives.

### *Citizen jury*

A citizen jury is a relatively small group of people, maybe 6–12, which is selected to decide on a community issue. We can say it is a form of a special committee, with more power. Power for one issue is delegated. A problem-solving or conflict-resolution function will not materialize if there is strong doubt about jury selection. Juries need to work according to strict timelines and produce well-argued and specified reports. A citizen jury can break open an impasse, force a shift in perspective on a local issue, but this only works if the decision is respected, if the jury embodies internal diversity in perspectives and has the power to shift policy in a meaningful way.

### *Temporary uses*

Using spaces or buildings for just a little while, either regularly, or just once, can be practical, as less investment is needed, as experiments can be accommodated, as spaces can be shared, and unexpected encounters facilitated. Temporary urbanism and pop-up design advocates argue that temporary, even ephemeral, uses can make a community more resilient, by allowing for the combination of more uses, addressing more demands, sparking creativity, and dealing with uncertainty. Temporary uses can normalize a diversity of stories about the places and create a mechanism for not only new activities but also new stories to enter.

### *Spotting the futures of the past*

This method belongs to the analysis of governance paths; yet, it is also useful in strategizing, even where not much self-analysis took place. Many strategies do not have much effect, and if so, not the intended effect, while others – though not well known or immediately understood as strategic in nature – may have quietly put the community on a different course. Collecting and assessing the old, imagined futures – the tools used to bring them closer, and their actual effect – reminds the community of what worked and what didn't, and to what extent various assets were discussed, implied, utilized, or overlooked.

### *Visioning and participatory design*

Visioning, as the phase between self-analysis and strategy, can take place in more and less participatory settings. Public discussions, open houses, and debate, in other words verbal genres, are common. Participatory design receives less attention. The hope, however, is that one narrative about the future emerges, often embodied in one spatial form. Design can help to solve several problems at the same time, discover and create qualities more easily, and it helps to make the resulting vision more comprehensible. The flipside can be that spatial form and maybe beauty carry the participants away, while underlying issues and narratives are not discussed thoroughly. Participatory design sometimes slides into design by committee, where a wish list is turned into a series of disparate elements, rather than a well-organized space that furthers a narrative and creates value. In each step of discussion, people with design skills can enter the fray and point out how ideas can translate into space and how spatial forms make sense or not.

# Index

Note: **Bold** page numbers refer to tables and *italic* page numbers refer to figures.

For Product Safety Concerns and Information please contact our EU
representative GPSR@taylorandfrancis.com
Taylor & Francis Verlag GmbH, Kaufingerstraße 24, 80331 München, Germany

www.ingramcontent.com/pod-product-compliance
Ingram Content Group UK Ltd.
Pitfield, Milton Keynes, MK11 3LW, UK
UKHW022157190926
475702UK00003B/69

* 9 7 8 1 0 4 1 0 2 6 3 2 7 *